UNCOMMON VALOR SERIES

Secret War Against Japan

THE ALLIED INTELLIGENCE BUREAU IN WORLD WAR II

COLONEL ALLISON IND, U.S. ARMY
with a new Preface by Steve W. Chadde

To
"Connie" and "M.J.F."

Secret War Against Japan
The Allied Intelligence Bureau in World War II

Colonel Allison W. Ind
United States Army

Maps by Donald Pitcher

with a new Preface by Steve W. Chadde

Preface copyright © 2014 by Steve W. Chadde
Printed in the United States of America.

ISBN: 978-1951682750

The Secret War Against Japan was first published in 1958 as *Allied Intelligence Bureau: Our Secret Weapon in the War Against Japan*, by Allison W. Ind. The original book is now in the public domain.

TYPEFACE:: Edita 10/13

Preface

THE SECRET WAR AGAINST JAPAN, by Colonel Allison W. Ind of the United States Army, is the dramatic accounting of the principal activities of the AIB–the Allied Intelligence Bureau–a group of British, Australian, Dutch, and American clandestine services formed in June 1942 during World War II. The AIB was under the command of Australian Colonel C.G. Roberts who reported directly to Major-General Charles A. Willoughby, General Douglas MacArthur's Chief Intelligence Officer for the South West Pacific area (SWPA). Colonel Ind served as Deputy Controller for the organization. Post-war, the AIB was disbanded and Colonel Ind served as a senior staff member on MacArthur's post-war occupation force in Japan.

The AIB performed roughly the same functions as the Office of Strategic Services (OSS, predecessor of the Central Intelligence Agency). The AIB was divided into four sections, A, B, C and D, with each focused on a specific area of operation:

Section A: consisted of Special Operations Australia (SOA), formed in March 1942 with assistance of the British Special Operations Executive. SOE was staffed with saboteurs, commandos and spies whose job it was to collect information and conduct commando operations. Specifically, Section A was charged with: harassing enemy lines of communications, attacking enemy shipping and small craft in harbors and rivers, organizing local resistance, establishing means of secret communications, distributing propaganda, and directing sabotage. In 1943, SOA was restructured and became known as Services Reconnaissance Department.

Section B: (Secret Intelligence Australia, SIA) was charged with signal interception and code-breaking. SIA was an offshoot of the British Secret Intelligence Service (MI6).

Section C: The Coast Watch Organization or Combined Field Intelligence Service, tasked with gathering field intelligence via coastwatchers, natives, and civilians. The coastwatcher network evolved from the pre-war Naval coastwatching system. The organization and activities of the coastwatchers is described by Commander Eric Feldt (who directed their operations) in his book *The Coastwatchers* (1946, and republished as part of the *Uncommon Valor* Series by Steve Chadde in 2014).

Section D: Far Eastern Liaison Office (FELO) or Military Propaganda Section; preparation of propaganda material useful to the other sections and initially for dissemination by them.

The first portion of *Secret War Against the Japanese* is devoted to the work of the Australian coastwatchers in the southwest Pacific islands. The members of this service performed a difficult, dangerous task in isolation and typically behind enemy lines in hostile territory.

The next major portion of the book details operations in the Philippines and the activities of Filipino guerrilla units.

The final section of the book, *The Commandos*, describes operations of Section A, the "Services Reconnaissance Department." They were a small group of able, courageous and experienced British saboteurs. The story is told of how they attempted to prove their worth by a mock demonstration of planting dummy limpet mines on allied shipping in Townsville harbor. However, this action was looked upon unfavorably by some higher-ups, who threatened to expel the group from the region. Eventually, they were permitted to destroy shipping in Singapore harbor, some 3,000 miles away, with notable success.

In hindsight, perhaps the A.I.B. was most effective in two areas: (1) deploying intelligence agents in places where they could gather information of immediate value to their commanders, and (2) assisting guerilla groups and conducting guerilla operations in enemy-held areas. An excellent example of the first type of activity was provided by the coastwatchers; the second was most successful in Bougainville, New Britain, the mainland of Australian New Guinea, the Philippines, and in the mountains of Borneo.

UNCOMMON VALOR SERIES

Secret War Against Japan is part of a series entitled *Uncommon Valor*, taken from the quote by Admiral Chester W. Nimitz, U.S. Navy:

"Uncommon valor was a common virtue,"

referring to the hard-won victory by U.S. Marines on Iwo Jima. The intent of the series is to keep alive a number of largely forgotten books, written by men and women who survived extreme hardship and deprivation during immensely trying historical times.

Steve W. Chadde
SERIES EDITOR

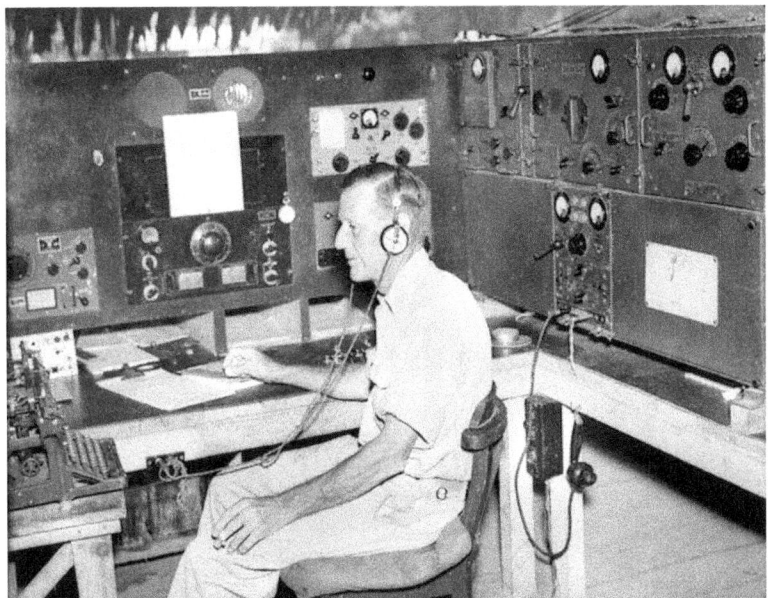

Lae, New Guinea. 1945. Chief Petty Officer Telegraphist J.T. Allan, Royal Australian Naval Volunteer Reserve, Control Wireless Station, Allied Intelligence Bureau, New Guinea Headquarters.

Australian soldiers and native scouts in the Solomon Islands.

Coastwatcher meeting with U.S. Marines officers, Guadalcanal, 1942.

Foreword and Acknowledgment

NAPOLEON ESTIMATED THAT "a spy in the right place is worth twenty thousand troops."

Initially in the war against Japan, MacArthur had few troops because of the holocaust raging over most of the globe; it was the task of the newly created Allied Intelligence Bureau to get spies into the right places, not only amid the bewildering island network of the Philippine Archipelago, but in the whole vast area from Singapore throughout the elongated tail of the Netherlands East Indies into New Guinea, the Admiralties, New Britain, and down into the Solomons. That area could be superimposed over the greater part of the United States and Canada.

The Bureau was one of the several intelligence tools originated and utilized by GHQ, Southwest Pacific Area, first for the collection of information vital to the integrity of the Command as a fighting machine, and then for the purposeful prosecution of the war by that Command. The history of each of those tools is a story in itself. This is "AIB's" story. But similarly, the sequences presented here can in fact be only highlights of a complicated picture involving at the end several thousand individuals performing allocated assignments. Of these a total of 164 were known to have lost their lives while the fate of 178 others remained a mystery. Seventy-five were listed as captured. A measure of the esteem in which these men, living and dead, have been held by the Allied governments concerned may be seen in the fact that approximately 170 decorations were awarded to all ranks and grades.

Official statistics credit the Bureau with a total of 264 missions, exclusive of those operated into the Philippines from mid-1943 onward under the semi-autonomous Philippine Regional Section springing from the original Philippine Special Section of the Bureau. The greatly stepped-up, purely AIB, effort mainly against the enemy-held Celebes islands and Borneo in the first seven months of 1945, mostly by the Services Reconnaissance Department of the Bureau (SRD), alone accounted for 155 sorties, the majority by aircraft. More than 300,000 pounds of supplies went in to agents, guerrillas, and isolated civilians. While combat is not a function of clandestine intelligence units, commando and other paramilitary operations of the Bureau accounted for more than 7,000 enemy killed and 150 captured, while still another 950 enemy surrendered in consequence of propaganda efforts carried on largely by the associated Far Eastern Liaison Office (FELO). Last, but certainly not least, on the humanitarian side were the more than 1,000 individuals of all services rescued by AIB units throughout the area of operations.

Just as the total effort in discharging the total intelligence responsibility is to be shared by all the intelligence organizations concerned, both those under and associated with the "G-2" of GHQ and those beyond SWPA, and just as the record of the specific AIB achievement is to be shared by all who participated in one capacity or another, similarly credit for the production of this single volume must be shared among numerous individuals, military and civilian; without their assistance, some over a period of more than a decade in the collection and analysis of data and the review of successive manuscripts, it is doubtful if the work could have been accomplished.

It is desired, therefore, to mention at least the following: Major General C. A. Willoughby, retired; Brigadier K. A. Willis, retired; Brigadier General Harry O. Paxon, retired; Colonel C. S. Myers, retired; Master Sergeant Juan Dahilig; Professor L. H. Conrad, Mary Jane Finke; R. C. Galang; Edith D. Johnson; Joan Corrigan; Anastasia Stamathis; Lila Beehler; and then, throughout, my patient, hard-typing wife.

Contents

Part 1	POINT OF NO RETREAT	1
Part 2	THE SOLOMONS	13
	The Coast Watchers 15	
Part 3	NEW BRITAIN AND NEW GUINEA	57
	New Britain Interlude 59	
	Buna 70	
	New Britain Toll 77	
	Gazelle Necklace 83	
	Fateful Hollandia 91	
Part 4	THE PHILIPPINES	101
	"Planet" Project 103	
	The "Man-Who-Walks-Like-a-Ghost" 129	
	"Carabao Boy" 137	
	Chic Parsons, "The Artful Dodger" 140	
	Sulu Sharpshooter 166	
	Philippine Snatch 174	
	Mindanao Mender 180	
	The Charlie Smith Way 196	
	The Captured Plans 202	
Part 5	THE COMMANDOS	213
	Limpets for Singapore 215	
	Tahoelandang 230	
	Sultan's Ransom 240	
Part 6	FINALE	253
	The Unspoken 255	
	Finale 257	

Part I
POINT OF NO RETREAT

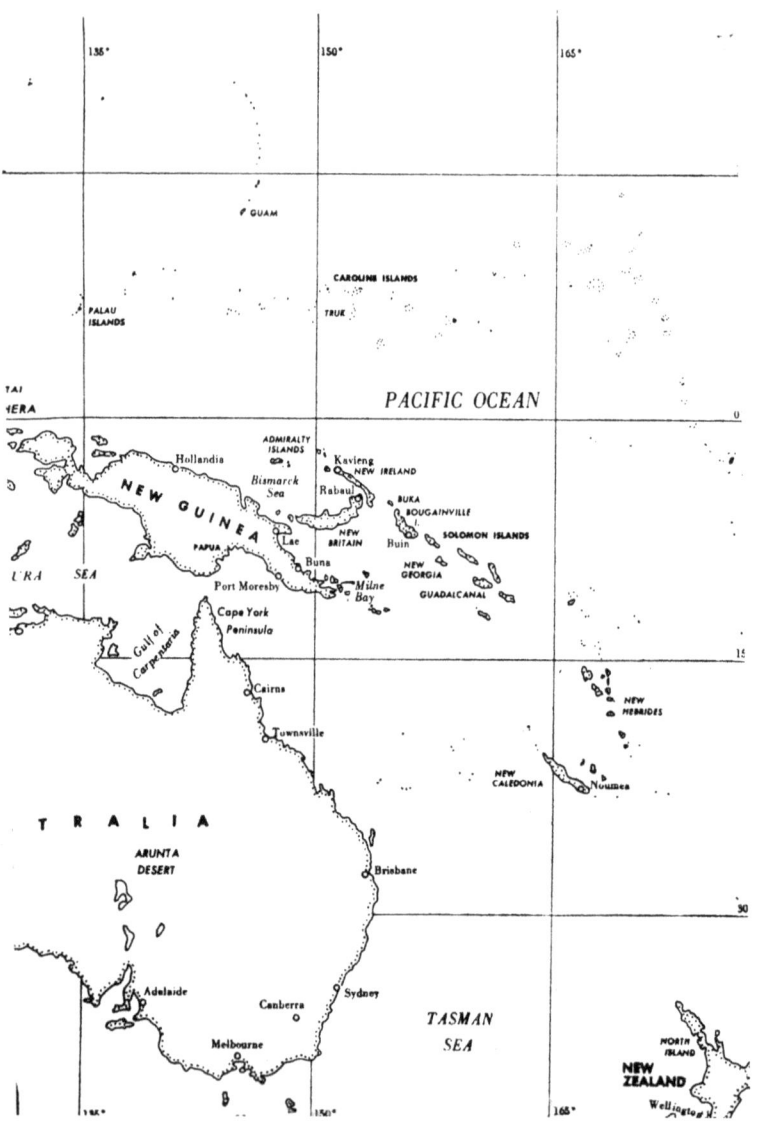

MELBOURNE, AUSTRALIA, ON THE SOUTHERNMOST coast of the far-south continent. Beyond, only the relatively small dab of land known as Tasmania. And then the wilderness of the South Pacific Ocean congealing into the white reaches of Antarctica. . . .

Melbourne, then, in early 1942; this far had the remnants of those defending the Western way of life in the Orient been driven from Singapore's sea-girt fortress, from the Netherlands East Indies, and the Great East, and from the stricken archipelago of the Philippines.

The point of no retreat.

Northward of Australia, a broad arc over the globe's surface was a sinister wartime secret. Forces moved there, gathered, and moved again. The evil of their intent and their implacability went before them like an infection. They knew no barriers, certainly none that we in the Philippines, the Chinese on the Asia mainland, the British in Malaya, or the Dutch in the Indies had been able to erect against them. One by one the thin, uncertain lines of communication between General MacArthur's reconstituted headquarters in Melbourne and the tortured Philippines three thousand miles to the north were fading into silence.

Singapore was mute. Batavia, Bandoeng, Soerabaja. . . . Rabaul in New Britain. New Ireland and the Admiralties. Paralysis was creeping down the Solomons, too. Cryptic signals still were to be picked up from a handful of Australian Coast Watchers who miraculously had survived the engulfing flood of invasion and now were isolated radio voices far back of the enemy lines of advance; other brief whispers came through the ether from ghost ships that hit and ran and hit again far in some other direction to try to deceive the enemy into a belief of strength when there was no strength after Pearl Harbor; from the few air strips still in Allied hands young pilots hardly more than out of flying school, yet suddenly old in the ways of war, flew daring missions without escort, always at the ragged edge of endurance, to bomb, strafe, and fight, and to collect any bit of military information that would help.

On April 18, 1942, General Douglas MacArthur had formally assumed his newly constituted command, Southwest Pacific Area, or "SWPA." This gave him military authority there over all the army, navy, and air elements of the Americans, the Australians, British, Dutch, and any other remnant of the Allied nations rolled back by the Japanese since the initial blows that had fallen simultaneously at Pearl Harbor, Hong Kong, Malaya, and the Philippines on December 7, 1941. The vast area of command responsibility ran westward to Singapore, ended at the Japanese Ryukus north of the Philippines, and to the east shared the remainder of the South Pacific with Admiral Chester Nimitz. It was a tremendously impressive command–on paper. There was only one thing wrong: the Japanese high command had not concurred in its planning; in the driver's seat of the most ruthless Asiatic military machine since Genghis Khan, they were in a superb position not only to contest its implementation but to smash it stillborn.

A month before GHQ SWPA was created, war-battered Flying Fortresses had carried MacArthur and his staff out of the beleaguered Philippines on orders from Washington and put their wheels down on the red, hot soil of northwestern Australia. And now the needs of the new command were as endless as they were adamant, for what MacArthur found in Australia was a monumental inadequacy to face even a fraction of the force he knew Tokyo could bring against him. He needed whole convoys of troops, shiploads of vehicles, radios, guns, ammunition, aircraft, and fleets of ships themselves. He needed parts, parts, parts. . . . But of all the needs, none was more immediate than that of tactical intelligence information about the enemy. The new Commander in Chief believed that the enemy's strategy was clear enough; it was not a question of "what," but rather "when," exactly "where," and "how moving."

First and foremost, as he saw the broad picture, the Japanese high command must see to it that at any cost Australia was not permitted to be developed as an American base. Unless severed at once, the vulnerable umbilical cord stretching thousands of miles across open sea to the arsenal of mainland United States would nourish that base into a giant's strength. Accordingly, that vital link had to be destroyed now. Since the loss of

Malaya and the Indies by the Allies already had isolated the British in far-off Ceylon and India, the parting of the two remaining principal obstacles to Japanese control of the Pacific world would insure each being dealt with separately. That should not prove difficult with American naval power a shambles at Pearl Harbor and Allied air strength cut to a few riddled, disorganized squadrons to cover a fourth of the earth's surface.

Severing this umbilicus, then, could be reduced to a simple tactical problem of gaining immediate control of a few key geographical spots north and northeast of Australia. That huge southern continent, as large as all of mainland United States, was ringed in those directions and to the east by a series of islands forming a great, broken double chain. There was New Guinea to the north and its satellite islands trailing southeastward from Milne Bay. Beyond this was the New Ireland-Solomons-New Hebrides chain, also stretching southeast. Bridging the two chains at the northerly end was the sizable single island of New Britain. Every ship to and from America would have to penetrate one or both of these island screens, or sail far to the south to New Zealand. But if Japan controlled the island screen and based naval and air raiders there, even the long southern approaches would be impossibly costly to slow Allied convoys either totally unescorted or given such scant protection as the Allies could scrape together. In any case, neither convoy nor escort could be expected to live long.

On the eighth floor of the tall bank building at 121 Collins Street in Melbourne, MacArthur conferred with his chief of intelligence, tall, handsome Colonel (later Major General) Charles Willoughby. That personification of genius and vitriol, Prussian drillmaster and lecture-room academician, had just consulted his huge situation map, had snatched from us in the G2 Section the latest teletype reports, and had conned the newest estimates from the Allied capitals. Brief reports from those magnificent Coast Watchers, from the few defiant ships, and from the individual pilots who performed as entire squadrons, provided new information to strengthen the Commander in Chiefs concept of the enemy's strategy: already he was moving down into the Solomons, he had occupied Buka Passage, Faisi, and Bougainville. He had been sighted in the New Hebrides, farther south. Down there was the important French base of Noumea, on New Caledonia island.

As the two men studied the situation it appeared increasingly that to control the island screen the enemy would need to anchor his effort at two extremes: Port Moresby on the under belly of New Guinea in the north, and Noumea, which lay due east of the upper third of Australia.

Willoughby described the scene to us later, the general pacing back and forth as he talked, striding with the energy of a man quite unconcerned that at the end of each five paces hard walls compelled him to about-face and stride the other way. He leaned forward as he went and his chin was outthrust, his hands tightly interlocked behind him. It was a familiar picture to Willoughby and the others of us who had seen him on Corregidor where the restraining walls of Malinta Tunnel laterals were uncounted tons of solid black rock. The Commander in Chief suddenly faced Willoughby and mentioned a name that was destined shortly to imprint itself forever on the pages of American history: Guadalcanal.

The enemy would seize some portion, probably the central area, soon; his creep down Buka Passage and Bougainville into the New Georgia chain presaged it. Obviously, unless he could be prevented from taking and developing a base in central Guadalcanal, where the kunai grass plains along the eastern coast would prove rewarding to diligent airfield construction efforts, his air and sea raiders soon would be enabled to range as far north and as far south as they wished in order to convert the Coral Sea and the Solomons Sea into Japanese lakes. He would have progressed far toward the absolute isolation of Australia, and America would likely be compelled to fall back upon Hawaii.

Ever associated with the threat to the communication link with America was the equally black possibility of the early invasion of Australia itself from such an occupied island screen, including New Guinea. Or the attack might even come sweeping down from the occupied Philippines, or in from the occupied Netherlands East Indies. (Already there had been devastating bombing attacks on Australia's northwest and west coasts; on the east coast an Australian military defense plan drawn up prior to the establishment of GHQ SWPA envisioned the loss of everything north of a line drawn across the country from some point midway between Brisbane and Sydney; a tremendous stand would be made to preserve the steel works at New Castle, just north of Sydney, but Brisbane would be gone.)

First things first. The threatened Solomons, then ...

GHQ had no troops to send northeastward to meet any threat, and very little air power. Obviously, the responsibility for meeting it would be up to Admiral Robert Ghormley, Nimitz's wing man assigned to protect the South Pacific (actually the line between SWPA and "COMSOUPAC"— the Navy's South Pacific Command—bisected in the Solomons). Ghormley would need *intelligence* to know how to husband his all-too-slender naval resources in order to hit suddenly and crushingly and get out of it to hit again. After they had met and beaten back this threat, said MacArthur, they could breathe, then they could dig their heels in and prepare to take that first step back to the Indies, back to the Philippines ...

". . . to win this war, Charles."

Win the war! At a time when survival itself was a day-to-day goal and some of us who had come down to Melbourne from Bataan and Corregidor had not even unpacked the emergency rations in our knapsacks because—well, frankly, it seemed to people who had been on the run since the first bombs crashed down December 7 that such an act of confidence would constitute a foolhardy tempting of fate.

Willoughby, who had been told to "land on your feet running" gave every appearance of having done so in his tremendous efforts to overcome the dearth of intelligence information. In rapid order he planned and secured approvals for half-a-dozen projects that were to be of great importance. One was Allied Geographical Section, to produce detailed terrain studies of vast tropical areas that had not even been names to the great mass of American officers and men. In an appallingly short time they would have to fight for them, live in them, and die in them. Another was Allied Translator and Interpreter Section, which before the war was over would produce solid bales of vital information from captured enemy documents, diaries, and interrogations of prisoners. He created and instilled efficiency into organizations for processing the strategic information that came from all over the Allied world as well as the meager bits and pieces that came from the immediate tactical front.

This tactical front was immense; fronts normally could be measured in thousands of yards; this one extended over thousands of miles. Compared with the compressed areas of the European theater, the enormous reaches involved in even the simplest operations in the Pacific staggered the imagination. The tactical and strategical areas of responsibility for GHQ SWPA could quite effortlessly be transposed over the whole of the United States and Canada with the east and west extremes well out into the Pacific and Atlantic oceans. With GHQ in Melbourne, the transportation span required to reach the area of threat discussed by the Commander in Chief and Willoughby that day can be envisioned by transplanting GHQ to New Orleans and considering ways and means of stopping an enemy expansion in the lower Hudson Bay region of Canada. Or if the Indies or Singapore were under study, it would be necessary to think of points far off the west coast of Oregon. Yet these were the *immediate* areas of concern; to mention the Philippines from which we had been routed was to consider points in northwest Canada and the Yukon region of Alaska. And all this with practically no long-range air units and, at that time, only half-a-dozen operational submarines that would have to be based at a point similar to San Diego, California, on our transposed situation map. Later, submarines would work out of Brisbane, Australia, or, let us say, out of Norfolk, Virginia—still a devastating distance in hostile miles to Hudson Bay or Alaska.

Obviously all ordinary means of reconnaissance for intelligence information either were hopelessly inadequate or would be consistently dependable only as the Allies won control of ever-advancing bases. This, then, was a job for ships and aircraft—but more than for either at this stage, it was a job for *spies on the ground*, sending in their reports from far, secret places by special radio equipment taken in by them. Before this could be accomplished, a coordinating organization would be necessary for the training of new agents, for the support of those already out, for communications, supplies, and equipment, transportation—a thousand things.

The foundation bricks for such a structure already existed—or, more properly, the initial tools for the secret collection of intelligence in denied areas were available in Australia by the time the wheels of the B-17's set down there in March 1942. My own acquaintanceship with these developments dated from mid-May, when I transferred out of Air Corps intelligence and into the incomparable "ulcer factory" of Willoughby's G2 Section, GHQ.

It was on a typical Melbourne winter day, cold, wet, gloomy, rendered the more depressing by the pale filter of light through the opaque windows of 121 Collins Street, that I reported to Willoughby.

"Go through these." He handed me a bulk of top-secret files. His left eyebrow was raised in a characteristic manner I had first observed in Manila. Those close to him had learned to interpret and evaluate that eyebrow; it was a barometer indicative of many things besides storm—but certainly storm. It might imply: *this is important*. It did this time. He explained that these temperamental, unconventional tools of warfare—saboteurs, secret agents, the Coast Watchers, commandoes, and so on—comprising the burden of the secret files—had been sources of worry and harassment to the Australian commander in chief, General Thomas Blarney, a silver-haired soldier of the orthodox school. MacArthur had agreed to take them all over, Australian, British, Dutch, and a few others, and Blarney had breathed a sigh of relief. We would have to find a way to operate and control them to best advantage in some plan to be jointly supported by the American, Australian, and Dutch governments. Washington had suggested to GHQ that it accept the services of General "Wild Bill" Donovan's OSS, or Office of Strategic Services, to do this clandestine work. But while MacArthur was cordial enough to the OSS founder and chief, and gladly accepted his offer of the services of the brilliant late Dr. Joseph R. Hayden, former vice-governor of the Philippines, as a confidential advisor, he declined OSS. General MacArthur felt that the various unorthodox units he was taking over from General Blarney and the Dutch might submit to a certain amount of control from him there on the spot, but he was convinced that an attempt at domination by or absorption into another intelligence unit based in Washington would prove to be unworkable. Besides, he required something immediately responsive to his requirements and command.

As stated, the files revealed that the "tools" were not exclusively for gathering intelligence information. One file dealt with an outfit that specialized in every phase of sabotage and silent killing. "Special Operations Australia" was its classified name. It was a branch of the world organization finely culled and sharpened in England and called Special Operations England, or "SOE." Factories, ships, power plants, arsenals–persons–anything and anyone valuable to the enemy any place, were "SOE" targets. In Melbourne SOE's headquarters was at "Airlie," a cold gray house behind a cold gray stone wall in the fashionable Toorak section of the city. "Airlie" was security tight and was personneled by British, Dutch, and Australian specialists. Singularly one-minded was its chief, Lieutenant Colonel G. S. Mott of the British forces. Moody, dark, saturnine, quick to anger and quick to act, he seemed to burn with deep inward resentment because the Japanese had routed him out of Burma and Java in turn. Yet his enmity was impersonal; he wanted only to turn his people loose against the common enemy. Before the end his "Services Reconnaissance Department," or SRD, as was its eventual open name under successors to Mott, would cover itself with glory, both as a sabotage unit and as a spy outfit.

A second file dealt with another British organization. This one was as pedigreed as SRD was new. The antecedental line went back to the sixteenth century. The communication channel of this unit still was direct from its own radio towers in Australia to Number 10 Downing

Street. London-born Commander—Wentworth, we shall call him—was director, and a shrewd, capable, imaginative dangerous man he was. When all other AIB efforts would founder in a prolonged welter of agent casualties in Netherlands East Indies, his radio monitors at war's end would be copying highly revealing cryptograms from spy operators under the very noses of the Japanese in Java.

The subject of a third file, Netherlands Indies Forces Intelligence Service headquarters, or "NEFIS," was coming into existence in a dark-paneled house on Domain Road, Melbourne. There was not much left after the rout from the Indies and Dutch New Guinea: a handful of loyal Indonesians, a score of Netherlands naval and army officers, and a lot of stubborn Dutch will. The Dutch govemment-in-exile in London sent money, some old submarines, and authority to participate in the new effort to the limit of their resources. This they did.

Fourth, there was the nucleus of what was to burgeon into an efficient propaganda service under the direction of Lieutenant Commander J. C. R. Proud of the Australian forces. Early in the history of the new unit, however, it would become evident that this specialized branch could operate more effectively under Australian control at Canberra, the Commonwealth capital.

The Fifth Division was already doing a superb job. These were the Australian "Coast Watchers." I had encountered them first in a prewar survey of intelligence assets in Australia. Since the first fateful days they, or those that still were alive, had been reporting vital ship and air sightings and the movements of Japanese land forces in the coastal areas of the islands to the north and east.

In June, when the Australian winter season was a damp, raw reality, Willoughby's executive officer summoned me. The late Colonel Van S. Merle-Smith of the prominent New York family spoke in cool, deliberate words of the need to draw up an official document to activate and operate what we agreed would be called "Allied Intelligence Bureau." The "controller" would be the current director of Military Intelligence for the Australian Army,

Colonel C. G. Roberts. I would fill the slots of deputy controller and finance officer. The decision to make an Australian the controller was based on both diplomacy and foresight. The cooperation of the Australians would be encouraged and when the time should come that the United States forces would have won their way back well north of the equator on their way to Manila and Tokyo, it would facilitate returning the remaining sections of the organization to full Australian control in the south. With an American finance officer, GHQ still had indirect but vital control: without his approval, any proposed operation would die stillborn of financial anemia. In Roberts GHQ had found a man of integrity, tremendous energy, and fearless loyalty.

Throughout the month, Roberts and I met with Merle-Smith to formulate the directive establishing "AIB." It was published in orders July 6, 1942. The Bureau was to "obtain and report information of the enemy in the Southwest Pacific Area, and in addition, where practical... weaken the enemy by sabotage and destruction of morale, and... render aid and assistance to local efforts in the same end in enemy-occupied territories."

It was succinct, it was ambitious. Considering the general Allied positions, it was in some ways a colossal presumption.

Space on the fifth floor of 121 Collins Street was assigned to us. Like the G2 Section, AIB itself was to "land on its feet running." Furniture, files, telephones, and short-tempered demands for immediate espionage plans involving the threatened northeast area arrived simultaneously. One of our first and most important visitors was a slender individual in the neat, dark uniform and spotless white cap of the Royal Australian Navy. His blue-gray eyes had a youthful twinkle; they and his smooth, unlined face belied the iron gray of the hair. This was Lieutenant Commander Eric Feldt, charged by the Australian director of Naval Intelligence with the expansion and development of the Australian Coast Watcher system. He unlocked a brief case and spread charts, maps, and statistics on the barely dusted-off tables. There was an upward tilt to his lips and an upward tilt to his words, but there was no nonsense about him. So it was that in a surprisingly short time we had made a compilation of names, check marks, and neat columns of notations that constituted Allied Intelligence Project No. 1A: the collection of all possible information about the enemy on the ground, in the air, and on the seas surrounding Guadalcanal.

That was the first project. In rapid succession would be others to cover Bougainville and the rest of the Solomons, New Ireland, New Britain, and New Guinea. Other projects directed by other men ultimately would aim for the Indies, Singapore, the islands of the Great East, and the Philippines themselves until, by the time Japanese officials were signing away their nation's sovereignty in surrender ceremonies aboard an American warship in Tokyo Bay some three years later, more than five thousand men and women would have been engaged directly or indirectly in hundreds of individual projects. Some of the efforts would prove to be tragic, many abortive despite the best that good men could give, some enormously fruitful—and all laced with the challenge of high adventure. Yet there was something more: a tenuous moral connective fiber throughout, a dedication by free men whose faith was made manifest in their actions during endless months of training, of routine grind, and in sudden action in far places. Capriciously, fate seemed to single out some for stellar roles. Two of these in the early days were W. J. "Jack" Read and Paul Edward Mason. On the day Lieutenant Commander Feldt spread his maps and charts out at AIB in Melbourne, those two Australians were at opposite ends of a mountainous, heat-bathed, tropical island in the Solomons, approximately twenty-five hundred airline miles to the northeast.

Part 2
THE SOLOMONS

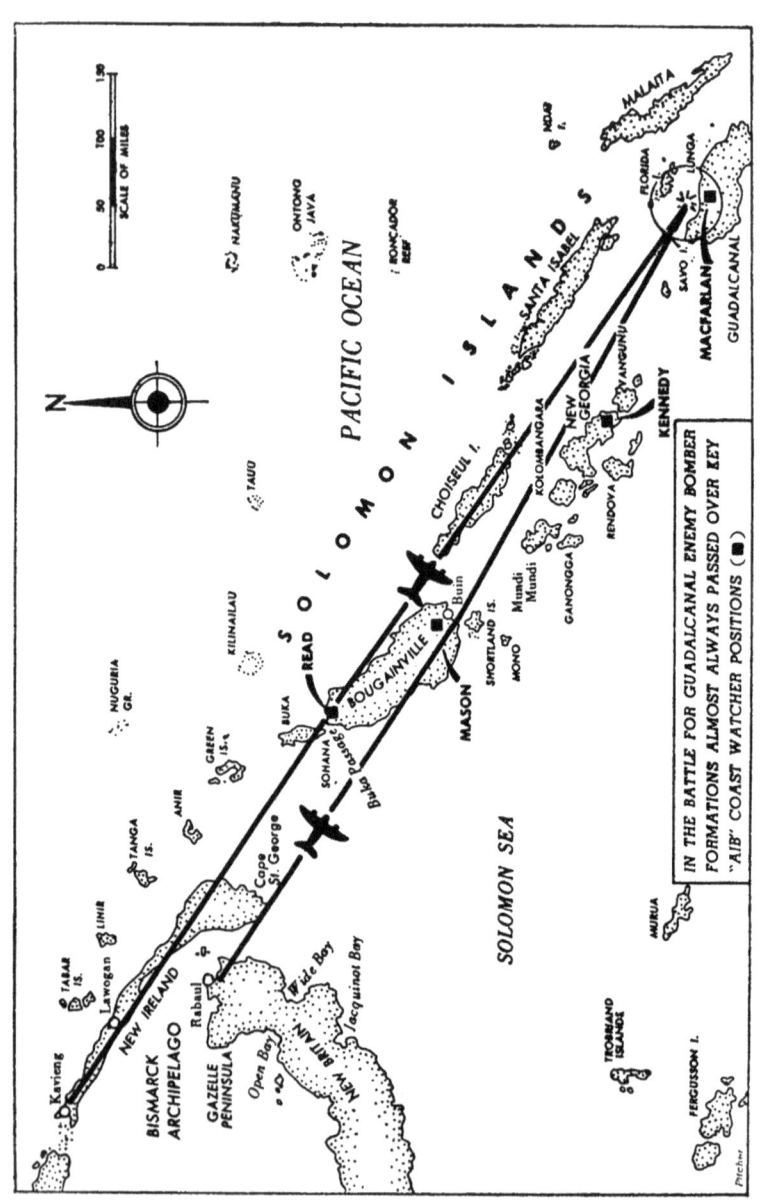

The Coast Watchers

IT WAS NO ACCIDENT that at the moment in history when their services would be most critically needed, Read and Mason, together with a priceless handful of others of their daring kind, were in the right places at the right times. Against the unproud record of the democracies for bland unpreparedness in the face of unmistakable warlike trumpetings there gleam a few bright exceptions. One of these was the Australian Coast Watcher organization which was bom years before the outbreak of the Japanese war at a time when to seek funds for military preparations was nothing short of heretical—and quite unrewarding. The first suggestions that Australia organize such a system had come as far back as 1919 from a Captain C. J. Clare of the Australian Navy. In consequence of World War I settlements, Australia was assigned mandates over the eastern half of New Guinea—Papua— and a large portion of the island screen that was to play such a vital role in 1942-43. The area Australia did not control was for the most part under the British Solomon Island Protectorate, except that New Caledonia was French. Not only for the sake of good colonial government communications but to provide an early warning of sea or air raiders in case of war, it seemed to Captain Clare advisable to establish a network composed chiefly of civil servants, the managers of copra plantations, and other "islanders" devoted to a remote way of life among the Melanesian native peoples, many of whom were very primitive. A genuine affection for the natives and a respect for their simple but solidly established rights were essential ingredients of the successful "islander." Fortunately indeed for the Coast Watchers of a later day, such factors were pivotal in the colonial policy of the government at Canberra—a stem but fair guardianship which in turn won the confidence and trust of the native. Without that, no watcher organization could have survived more than a few weeks at best when the chips were down, for in its essentials this specialized espionage was no different from other espionage activities: the degree of success would be directly proportional to the receptiveness of the host peoples involved.

A civilian, Mr. Walter Brooksbank, was primarily responsible for molding the foundations of the embryonic watcher organization, and incorporating into it the office of the director of Naval Intelligence in Melbourne. There, Commander R. B. M. Long, a stocky man with a cupid's-bow mouth and a steel-trap mind, added his official weight and skill. But it was a labor of love; World War II would be casting its unmistakable shadow before he would be able to generate even mild interest on the loftier levels of the political and military. He went ahead anyway, appointing Lieutenant Commander Eric Feldt of the Naval Reserve to get the detailed field work done.

Feldt did it in the far reaches where tropic seas creamed against white beaches or soughed in the mangrove swamps of lonely coasts with their spattering of latticed houses and atap huts amid ranked coconut plantations. He consulted and recruited among the civil servants, likely themselves "islanders" and probably also planters. Feldt talked their language, for he was an "islander." He was "right." They signed on. They heeded his suggestions of what to look for, how to utilize "flash codes" in reporting air and ship movements by radio to the net control station at Port Moresby— all at no salary. Moresby would relay to Townsville on the Queensland coast, and Townsville would push the messages farther south; in a short time Commander Long would add another "flag" on his situation map in the office on St. Kilda Road, Melbourne. It was this organization and these "flags" that a military mission from Manila, of which I was a member, studied with such interest in November 1941–a month before Pearl Harbor. We would shortly return to Manila and report that when war came—as we then knew was inevitable—there would be some splendid observers already in key spots.

So it was that as the first projects were formulated that day in the AIB office in Melbourne, Read and Mason were in such spots on Bougainville at the head of the Solomon chain. I became personally acquainted with Read and some of the others as time went on, and from their unadorned narratives filled out the cryptic reports they prepared.

As a result of enemy action during the last days of January 1942, Read already had become a "gypsy" without a permanent home. But he was within sight of what had been his home all too briefly at the northern tip of Bougainville. It was quite evident that the enemy landing parties had "taken over the lot," as he told me many months later. "With the glasses we could make out the whole of the Buka Passage area well enough, and especially the buildings on the little island of Sohana." Read had been colonial district officer on Sohana. He also had been a civilian Coast Watcher. It would be some weeks later that he would become a lieutenant in the Naval Reserve, a development resulting from Long's tireless efforts in Melbourne to give military status to Coast Watchers in order to prevent their being classified as civilian spies and shot out of hand by a capturing enemy.

Read had developed an efficient "posi," or position, high above Buka Passage. The fluted column of a gigantic rain tree gave him flank protection while affording an unobstructed view to the northwest. He had arranged a support for his binoculars. On this particular day in early 1942 he had been observing the Japanese on Sohana. The shore parties once more were engaged in looting and smashing on Sohana.

He slowly shifted the glass so that the field of vision crept across the indigo blue of the Passage itself into the shallower green water, and then up along the hard coral ribs of Buka Island as it lay opposite him.

There was no sign of activity on the shore line. Carefully he examined the atap fiber huts of the Melanesian natives of Buka village. The few huts that had been burned in the bombings prior to the landings by the enemy had been replaced. Everything looked peaceful enough there, but Read had no illusions about these natives; they were realists. Throughout history they always had been dominated by someone stronger. Until recently it had been Australian colonial government. Now it was the Japanese. There would be no real affiliation, not even a nominal one, but that village would serve strength—and it had seen the white man run out by the Japanese.

Read glanced at his own Melanesian "boys" with their glistening black bodies partly covered by travel-stained calico lava-lavas, or wraparounds. Was it dangerous wishful thinking or a sound appraisal based on his ten years' experience with Melanesians that induced him to place implicit faith in the trustworthiness of this little group? They had seen him prepare to abandon Sohana shortly after the two Royal Australian Air Force Catalinas that had been based at Soraken, just to the south, departed one day, their crews taking possessions that normally they left until their return from routine patrols. Then news of the fall of Rabaul on the north tip of New Britain had come in through Read's radio receiver. The jungle telegraph had got it, too. Provident and calculating, Read previously had reconnoitered for a position on the high basalt upthrust country of interior Bougainville that would enable him if necessary to "hole up" with supplies and maintain active observation; it simply never occurred to him to try to escape, even though he was convinced that the tide of invasion would sweep beyond him as it then was sweeping over the islands just to the northwest of him. Out there on Taber Island had been Coast Watcher C. L. Page, who shortly after the outbreak of war had broadcast the warning of a four-motor Japanese flying boat that had soared over his lonely outpost to take sights on Rabaul. Soon Page had been run out. Then Allen on Duke of York Island went silent, and Chambers on Anir abruptly stopped sending. C. C. Jervis of Nissan, just off to the northwest from Buka, quietly mentioned that a Japanese warship had hove to off his tiny atoll, which he had described as being "as flat as your hand and just as bare of places to hide in." Silence again. . . . But there were others that the enemy had not found, and because of this oversight, the enemy would pay. Read was wiry of build, rather than big or solid, but within him was a solidness of spirit, a steely determination, and a capacity for careful planning; it was altogether likely that in his own way he, himself, would extract payment.

His loyal native police sergeant, Yauwika, asked permission to use the field glasses. His voice was hoarse and his pidgin English a little hard to understand. Yauwika frowned fiercely as he put the glasses to his eyes and stared toward Buka village. Read thought he knew what Yauwika sought: through his own devious means and loyal native missionary workers Yauwika had instituted a native "pipeline" and along it detailed information flowed in a mysterious current; now he knew exactly which of those atap huts he could see with their fiber-woven roofs concealed Japanese supplies, and which were serving as barracks. Read's cryptic messages to the Royal Australian Air Force—RAAF—made good, explicit bombing objective data.

Read studied the stocky, black native police sergeant. He decided that Yauwika could be completely trusted—and probably those he worked with. Yet power impressed Yauwika, too, and Read had been both disappointed and secretly amazed at the effect news of the Battle of the Coral Sea had had upon the natives of Bougainville. Read had heard of Coral Sea both from news casts and official communiques. The jungle telegraph of the natives had been almost as prompt. The naval collision had occurred in May in consequence of the Japanese attempt to clear the way for an attack against Port Moresby, which the Allied high command had foreseen as part of the strategy to control the island screen. Read was able to glean that the victory had been a tactical one for the Allies in that the enemy effort had been beaten down. But in the fight the enemy had inflicted more damage than he had received. Just how the native assessment had arrived at the truth was a puzzle which Read ascribed to the white man's general ignorance of native shrewdness. In any event, Coral Sea had been disappointing in its propaganda effects—and Midway was too distant to have any effect at all.

In addition, Read was apprehensive that the jungle telegraph had scored another local beat. A Japanese force had since landed on Guadalcanal and was preparing an air base on the kunai grass plains. What was needed urgently to avoid deterioration of general native support was a spectacular Allied victory in the Solomons with the results heaped up for all to see. If it did not come soon, he could foresee trouble of another sort on Bougainville. Here and there on the island were groups of defenseless civilians and some of the missionary personnel, including Catholic sisters, who previously had declined to heed his advice—and that of others who had no illusions—either to go overseas to safer areas while they could, or at least head for the mountains and avoid contact with Japanese patrols led by cooperating coastal natives. All told there were perhaps a hundred. They had been recalcitrant but Read had no doubt they would be demanding a protection there would be no way of according them when the bulk of the coastal natives should really turn against them.

Beside him Sergeant Yauwika had grunted excitedly and pointed. Read wiped the moisture from his brow and squinted. Sohana was still quiet enough. But something was stirring in Buka village. Yauwika fired an explosive torrent in his own dialect and his black-skinned comrades rolled excited eyes. Read rapped Yauwika's arm for the glasses and peered.

The scene was too confused to enable him to get the exact pattern but apparently the Japanese once more were resorting to harshness, even brutality, in enforcing demands upon the villagers. The Solomon Islander was not long out of the Stone Age, and from there he had come direct into the Steel Age. But he had brought with him almost intact his superstitions, his tribal customs, and his curiously well-developed sense of property. Of the latter, his most potent single symbol was his pig.

Possession thereof marked him a man of means; possession accorded him bargaining position for all things in his social scheme, from food gardens or a seat in the village council to his fancy in a wife. Coral Sea had upset the Japanese supply communications and here at Buka a levy was being made upon native resources to make up for the fish and rice that had failed to arrive. Not content with a diet confined to tropical yams, taro, and sago, the enemy soldiers and sailors demanded pigs. They refused to penetrate far into the unknown jungles to hunt for them; the alternate was to commandeer the native's prize possession.

Read could see a body on the ground of the village street. Nearby an enemy wiped a bayonet. Two others carried off a kicking pig. Sullen natives opened a way for them, then closed in after them. A cloud of dust slowly subsided. This execution of a villager who had resisted doubtless was what Sergeant Yauwika had witnessed.

Read felt a grim satisfaction. Word of that kind of treatment would get around. Of course it *could* terrify the native into a kind of cooperation that would be dangerous to him, but with some military development favorable to the Allies the enemy would soon know what it meant to alienate the natives. Read's own meticulous adherence to a code of recognition of the small rights of the native in a white man's world was the key to their loyalty to him. There was one other related item of influence: the Solomon native's appreciation of money. Read had taken all the money from the office at Sohana and could and did pay the fair price for their services and goods. After all, it required a carrier force of from fifteen to thirty "boys" to carry the radio gear and all, to say nothing of essential scouts and runners.

While Read was reflecting on all this, there was a sudden poising of weapons. Someone was moving in the giant ferns to the left. Then Yauwika grunted. It was one of their own lookout boys. He was breathing hard. He spoke fast in his own dialect. Yauwika converted it into mangled pidgin. Three ships, warships, were approaching Buka Passage from the west. Read questioned him. Apparently a cruiser and two destroyers.

Read signaled and the party moved off over a barely discernible trail so that he could verify the report. If true, and if the ships anchored, they would comprise a target worthy of any bombing mission—or if not, simply the knowledge of their exact whereabouts would be of importance to Allied planners moving their invaluable pieces on the chessboard of the South Pacific. An hour later Read and his natives were squelching over sodden trails to the base. Just in time the confirmatory sighting had been made before the rain streamed down into a jungle already steaming with dank heat. Read clumped into a hut; he hated the rains. He motioned for Tamti, a native who had been taught wireless signaling. Tamti warmed up the tubes of the Australian-built "Teleradio," a splendid piece of equipment that originally had been designed to service Australian "outback" sheep stations and the "flying doctors" who administered to their needs in response to radio summons for help. While he worked over the three steel-gray cases of the combined receiver and transmitter, Read enciphered. He wondered if the signal would reach distant Port Moresby on New Guinea—nearly seven hundred airline miles. It would be another eight hundred from Port Moresby to Townsville where the Intelligence Center was located. It was another fifteen hundred to Melbourne. He was considering the unpredictable behavior of Hertzian waves in the ether. The present saturated sky was certainly bad in theory for transmission, yet already Tamti had registered acknowledgment from Port Moresby. Another time the sky could be clear as crystal and Moresby might be off the planet as far as their capability of reaching the control station was concerned. Moresby confirmed receipt of the data and signaled no more traffic.

Tamti moved aside and Read played with the vernier dial of the receiver. Familiar voices came to being, for now he was on the frequencies assigned to the Coast Watchers, and where distances were short, radio phone, or "voice," was often used. Brief, staccato barks for the most part, because, after all, a radio transmission was a voice for all to hear—and the enemy missed few of them, day or night. For instance, he had all-too-clearly heard the enormously foolish newscast out of Melbourne not long ago stating that Japanese warships had landed reconnaissance forces not too far from Read's position. Immediately the enemy was interested. No aircraft had sighted that landing. A Coast Watcher, then? They knew of only one man that could have radioed such a report, a really harmless old chap named Percy Good whom they had put on parole. Two days later they landed again and despite his protestations of innocence—for innocent he was, since he was not a Coast Watcher and the report had been filed elsewhere by one who was—summarily executed him.

In Australia prompt steps were taken to insure that henceforth no such revealing information would reach the air.

Read often spent a part of the night hours this way, although he had always to be sensitive to the drain of his batteries. The Teleradio was powered by heavy wet cells that had to be charged by a petrol engine charger, even heavier. Petrol itself was no small problem. There were caches...

Meanwhile GHQ had moved bodily from Melbourne eight hundred miles north to Brisbane, and with it the headquarters of AIB. In one of the big insurance buildings on Queen Street, exciting matters concerning the Solomons were occupying the high command. But the cards were being held very closely to official chests and in general very few who had no "need to know" possessed the knowledge that soon there would be launched the first Allied attempt, not only to halt the enemy drive down the Solomons, but eventually to reverse it. It would be a combined navy, air, and land assault to annul the Japanese landings at Lunga on Guadalcanal and other strategic points in the central Solomons. GHQ was playing a support role this time, for the show primarily would be one of COMSOU-PAC's with the American First Marine Division scheduled for the actual landings. The concentration was being effected in the New Hebrides with Noumea as the hub.

"I didn't know anything specific," Read said later. "But I became properly suspicious that something was up, just by listening to the ear bashing over the air."

Lieutenant (later Commander) Hugh Mackenzie was AIB's Coast Watcher coordinator stationed with COMSOUPAC headquarters in the Solomons. Read often tuned in to see what Hugh had to say. It was not much. But from the same quarter came many broad American accents. Mackenzie had helped the Marines to organize their own local watchers. "Not a doubt of it–the Yanks were zealous," commented Read. "But they were too ruddy voluble on the air. Obviously something was up."

Another whom Read monitored was D. S. Macfarlan. One of the best was Macfarlan, with a gift for accurate observation and a matching ability to describe what he saw. Later in Brisbane we were to benefit from these talents to get vivid word pictures of what occurred at Guadalcanal largely in consequence of the roles that Read and Mason soon were destined to play.

Assigned to Guadalcanal to watch the Japanese at close hand, Macfarlan had insisted on a grandstand seat to enable him to accomplish the task with aplomb. He fairly looked down into their rice bowls from a lofty, beautifully hidden spot on Gold Ridge. He reported every inch of progress on the air strip the enemy was building below him at Lunga. Just to the east of Lunga was Tulagi, the former center of British colonial administration for the area. The enemy held that place, too. Macfarlan's boys had been reporting in detail defense strong points, stores, and barracks. He encoded the data and sent them two ways: to Mackenzie for use by appreciative COMSOUPAC authorities and those who were to lead the assault, and to AIB via Port Moresby.

Other Coast Watchers hidden on Guadalcanal were contributing to this vital build-up of information. There were taciturn F. A. "Snowy" Rhoades and A. M. Andersen. Martin Clemens was farther north. And off on the dreary basalt upthrust of tiny Savo Island was another, frail elderly L. Schroeder. On one occasion while Read monitored, Macfarlan was reporting the estimated strength of a Japanese force north of the partly finished airfield. Read was amused. He knew Macfarlan's method: send in natives who would come back and tell him how long the mess line extended. It was as good as any and better than most, for natives were notoriously unable to estimate numbers of men. "You'd get anything from a thousand to twenty times that number," Read would explain. "The true count was probably five hundred."

Guadalcanal was a somewhat fatter island than most in the Solomons. To the north was a spattering of small islands comprising the New Georgia group, between it and Bougainville. Other Coast Watchers were there. And they were important, for while Read was the farthest northwest of the whole series of Watchers and therefore could give maximum warning of raids that he might detect coming south to hit American concentrations, enemy ships and aircraft *might* not touch within his eyesight. Then it would be up to Paul Mason, a hundred and thirty miles south of him on the opposite end of long, thin Bougainville, to warn. After Mason came the watchers in the New Georgia group, and finally those on Guadalcanal. Thus there was observer strength "in depth."

On one night while Read monitored, Paul Mason was acknowledging receipt of a successful air drop, containing desperately needed supplies. Read chuckled explosively. The black boys squatting on their haunches in the opposite corner of the hut rolled eyes whitely toward him. In turn they grinned. They appreciated it when the *kiap* was amused. He was a good *kiap*, stem and quick, but a proper "big fella." Read was thinking that poor Mason deserved a successful drop after all the trouble he had experienced.

Mason had been a planter on Bougainville. He probably was less like the conventional picture of a conquering military hero than anyone Read could imagine—quiet, round-faced, with a boyish complexion that went with the mild blue eyes behind round spectacles. But we were to learn that Mason would surprise you. When the going was rough, he was indeed a good man to have on your side. His determination was matched only by his unsuspected powers of endurance. The first drop that had been made to him went astray by not less than seventy miles! But Mason doggedly set out by bicycle and on foot to recover it. His reward after a bitter trek through some of the most uncomfortably fetid country in the South Pacific was exactly nothing. The situation was doubly distressing because, whereas Read had plenty of supplies cached, Mason had had no opportunity to save much from the plantation before the Japanese landed. Read had expressed his concern that Mason's natives might not appreciate his straitened circumstances. The second drop would alleviate the situation, at least.

Mason had taken up a position at Malabita Hill near Buin, on extreme southern Bougainville. As fate would have it, the area he could effectively overlook would become one of the major anchorages for the Japanese Southern Operations fleet.

The drops to Coast Watchers included more than merely the means to keep body and soul together. Feldt had experienced many years' service on isolated assignments and he knew very well the tremendous importance of mail from home. Great care was taken to insure inclusion of one or two pieces at least. So it had been that the last bomber to drum out of the night and lay a softly floating parachute onto Read's carefully selected drop zone had brought a letter from his wife in Melbourne. It had been a solid, homey sort of thing, just what he needed most, and it had told him all about the antics of their little daughter, Judy. (What it did not tell him was that Mrs. Read had experienced a serious accident; she had not wanted to worry him for she was confident that the results would not be lasting.) Reference to Judy had made Read smile. Doubtless both his wife and young Judy would be amused if they knew of the arrangement he had with Feldt that if he ever had to go on the air in plain language and wanted to confuse listeners other than Feldt, he was to sign his emergency messages "JER," Judy's initials.

Read snapped off the Teleradio until the next schedule with Port Moresby. It was time for kaikai, or food, and despite the downpour that went on as if it never intended to stop, the boys managed their cooking and Read had his plate of hot tinned meat, vegetables, and his mug of tea. He hitched himself farther from the drifting spume at the propped-up atap transom over the open window space and savored the brew.

In the next transmission Read reported that he had made a short reconnaissance to the coast. At one point he had been waited upon by a surly delegation of the civilians whom he had urged to take proper precautionary steps when the war was young and there remained time to act. They demanded that the RAAF make air drops of supplies to them. After all, they insisted, the government owed them protection and sustenance, too. Read reminded them that protection was something they had rejected by keeping themselves beyond its line of possibilities. He reminded them that there were exactly twenty-five Australian soldiers on Bougainville, a poorly equipped guard detachment of prewar days, in almost as bad a plight as they were themselves. As to sustenance, he imagined he could forecast RAAF's reception to the idea that it should risk its few operational bombers from their round-the-clock war responsibilities to deliver food to people who a short time ago had denounced the military as unnecessarily dictatorial. But that night Read relayed their request to Port Moresby. The next day he had confirmation of his forecast. The issue did not endear him to the group, nor it to him. Read was honest with himself; he was direct and honest with others. He had an idea he had not heard the end of them. He wondered about the Catholic nuns he knew to be still on the island with a Bishop Wade.

He continued to file his observer reports from Buka Passage. Every significant bit about the air strip, the village, or the ships that came and went and what they did while they were there swelled our intelligence file at AIB.

But one day toward the end of July he got a deeply encoded message from Feldt at AIB.

He was to stop reporting everything except emergency information of great importance. He would establish himself at the best point of vantage above Buka Passage and await developments. Meanwhile, if he did see something very important, especially in the air, he would report directly *in plain language.* His messages would be identified by the "JER" call sign.

He stood in the middle of the shack, the deciphered message in his hands, and he knew this meant action. He was laying pounds to sixpence that it would be on Guadalcanal to the southeast of him. A few minutes later he knew that Mason had received the same instructions from Feldt and that he would use the call sign "STO." Mason had married a girl whose family name was Stokie.

It was fully appreciated at AIB that broadcasting in plain language meant shouting out the locations of the two men. But speed in reporting was going to be of utmost importance; conceivably the success of the whole operation would rest on adequate warnings of enemy counterblows while the landing and support ships were bunched without maneuver room at the beaches. Against the "give away" was the fact that the two men were tiny needles in a very large haystack. Bougainville was a hundred and thirty miles long by thirty wide and what was not smothered in rain forests and jungle growth was upended in mountains. As long as the natives did not turn against them... Nevertheless, it was a very uncomfortable decision to have to make back in the safety of headquarters.

Read looked wryly around the shack. It had been a secure spot. He decided to make a cache of supplies in several places, of one or two of which only he would know the exact location; then, whatever befell him, he should have recourse to a few emergency supplies. He was to be thankful one day for this item of his foresight.

The next morning he put into execution his plan for erasing signs of the base preparatory to moving northwestward. He was seeking a better position that was low enough to avoid most of the mists but high enough to cover most of the vital area. The rains came down as they departed. This time Read did not mind so much. He wanted their tracks to be erased. He used trails unknown to all but his little party, or cut new ones. On the evening of August 7, 1942, he still was some distance from Buka Passage. Except that it had been a strenuous day, for him it had not been unusual. Since his teleradio had been stowed in many containers slung from poles supported fore and aft on the shoulders of carrier boys, he had been entirely unaware that in reality the day had witnessed a tremendously important military development—in some respects even more significant than either the naval battles of the Coral Sea or Midway, because it represented the first retaking of a land mass for a counteroffensive that was not to experience a major reversal until the war was won for the West. Read made camp where the rain forest was so dense that only directly above him was there open space. After a new downpour the stars showed there.

Paul Mason, however, had been aware of something unusual on that day. He had, in fact, played a part in connection with it as vital as any who participated, and doubtless because he had, many still were alive and safe who otherwise would have been victims of the air attack launched by some twenty-seven Japanese warplanes from Rabaul.

At dawn of that day the first massive combined-services counterassault of the war had been launched by the Allies. It was against Guadalcanal, Tulagi, and other nearby Japanese-held targets. By the time

Read had pitched camp for the night, the First Marine Division was dug in hard along the Japanese airfield at Lunga, had retaken Tulagi, and had established firm beachheads elsewhere. At dawn that day planes from United States carriers had screamed down on Tulagi through a mist that hid the approaching battle and transport fleets from sight and had bombed it heavily. Then the approaching warships had divided. One division had taken over from the bombers while another steered for Lunga and opened up on shore defenses there. The transports came in and released their battle-ready Marines. The assault boats rammed square steel snouts hard aground and the heavy ramps roared down. The enemy had been taken completely by surprise, and on Guadalcanal were unable to rally against the attack. The survivors fled into the jungle toward the mountains. It was almost the same at Tulagi, but there was determined resistance elsewhere.

The Allied command was under no illusions about the initial success. The enemy simply could not afford to allow the Allies a secure foothold in the South Pacific, and certainly not one they themselves required if they were to secure Port Moresby, already overdue, and cut the supply line from America. The American commanders hoped that before the enemy could reorganize or call for help, the heavy task of getting supplies ashore to support the troops for a prolonged period of what was sure to come could be accomplished. But manhandling thousands of tons of the materiel of modem war from transport to barge and barge to shore and shore to truck and truck to supply dump was a task to weary the stoutest backs, especially when the debilitating tropic heat wrapped the whole area in a sodden blanket. There had been little or no time to rehearse this phase of it. The work went slowly; the wonder of it was that the sweating, swearing men had managed to bully as much of the artillery, the ammunition, the construction materials, and the rations ashore as they had by the time Mason's signal reached the ships to the effect that over his hidden position at Malabita Hill on Bougainville at 1030 hours he had sighted

TWENTY-SEVEN BOMBERS HEADED SOUTHEAST

The message was signed "STO." Actually, the ships had not received it directly from Mason. He had sent originally on the "X" frequency for emergency traffic and had been copied at Tulagi, also aboard the Australian cruiser *Canberra*, which together with another Australian warship had cooperated with the American vessels that morning in the shelling of Guadalcanal. Port Moresby also had taken it in and immediately relayed to Townsville. Townsville relayed to powerful transmitters outside Canberra which in turn flashed it to Honolulu. The transoceanic station on Oahu stepped up to full power and sent the warning roaring back over the Pacific for all to hear. The whole series of relays had taken approximately twenty-five minutes, and the Japanese still were a considerable distance off, coming in under heavy bomb and fuel loads.

Mason later said that he was barely able to contain his excitement and that his eyes "felt very hot" behind his glasses as he constantly fingered his receiver dials for news. The air was full of strange signals and he discerned both the Morse key "fists" and voices of Americans. He knew then that "The Yanks were fair in it now!"

Mason could hear only confusedly, and he could see nothing. Macfarlan on Gold Ridge, Guadalcanal, constituted the "eyes" for this show, although some of the others caught thrilling episodes of it from their lairs.

Macfarlan had noted the orderly deployment of the American carriers well out to sea. He saw the transports weigh anchor and disperse in a disciplined pattern away from the crowded beachhead. A low throb of scores of aircraft engines being preflighted came from the distant flat tops. The drone swelled and sounded to him like "a far-off bullish roar." Presently he saw our planes rising, looking in the distance "like some kind of slow daylight tracers." In due time they found their proper stations and began a leisurely climb into the high places of the sky. The heavier American fighter planes were not so nimble as the Japanese Zeros who doubtless were escorting the bombers, but the Philippines had taught us that if the American planes had the advantage of height, the aerobatic superiority of the Zero was canceled out: our heavier armor and armament more than evened the score. At Guadalcanal, local warning only would not have enabled the American fighters to have climbed high. Mason gave them an hour and a half to do it.

The Japanese pilots were unprepared for the magnitude of the reception awaiting them from below and were disorganized by the wide deployment of the targets.

To Macfarlan the whole sky had suddenly become dappled with the black puffballs of ack-ack fire from the ships. Now he could see the Japanese bombers coming in on the final approach, but the intense fire disrupted their runs and they began to fall off. Bombs fell into the sea and far off to one side on the land. One bomber then another burst into flame and black smoke. Others sideslipped and the ack-ack from hastily erected shore guns took them under fire. The attack seemed to be breaking up before it was fairly launched. The Japanese veered more and sought to disengage. The smoke now made observation difficult and Macfarlan could catch only snatches of the big surprise. At just the time when the enemy planes could least afford to take on a new menace, a whole sky full of fury dropped upon them from the heights. Now both bombers and Zero escorts alike were caught in a storm of tracers and shells. Pilots who remembered Pearl Harbor, Singapore, Clark Field, Bataan, let them have it again and again.

Paul Mason saw only one bomber returning that afternoon. This was a paradox, for initial Allied claims listed only two shot down and two damaged. That was before the carriers deployed out to sea had entered their tallies. Possibly more than that single bomber seen by Mason got away, but history was to say that it was doubtful if it had any companions.

The next morning, August 8, Read broke camp early. The party was on the trail when shortly after 0800 hours the column stopped as one man. The thundering roar of many aircraft engines seemed to burst upon them. That could mean only that they were flying low. Automatically his boys melted into denser jungle. Read looked around desperately for some point of observation. But only his boys were equipped by dress and nature to shinny up the boles of trees. Then the sound exploded directly above him. To his surprise he found himself reacting automatically to training even as his boys had done when they "melted" into the jungle. He was counting aircraft which, as fate would have it, were flying across the only place in the whole vast canopy of the sky that he could observe—immediately over his head. They were no more than five hundred feet above him and he could discern not only that they were bombers but that they were torpedo bombers.

The din receded. He shouted orders. Two of his boys were well toward the tops of tall trees. The others reacted quickly and began to break out the wireless gear. There was no time to make for a more suitable transmitting area. While Read knew nothing about the Guadalcanal show, he was certain that eighteen heavily-laden enemy torpedo planes were not out on an incidental early-morning training mission so far from Rabaul on New Britain or Kavieng on New Ireland. He suspected these were from the latter place; the direction was right.

He shouted to that boy, helped this one, and heaved with the third. The transmitter, the receiver, the batteries, heavy and dangerous to handle. Up went a jury antenna.

Then a warning call from the treetops. More aircraft. As before, the sound burst upon them. This time Read counted twenty-two aircraft; some were Zeros.

The sweat blinded his eyes. He dashed it out and tested the power couplings. Everything seemed to go maddeningly slow. But not really. "Just the ruddy excitement, you know," as he said later. He calmed himself, sobered by the thought that at this time of day it would be a real chore to raise Port Moresby—and on "voice," too, for he did not have Tamti this time, worse luck. Voice had less range than Morse.

He called on voice. Then listened. There was only the hissing of the tube filaments in his ears. He tried again, adjusting as he spoke. But the transmitter was putting out a maximum signal; adjusting would do no good. There was not even a crack of static.

He banged his fist in exasperation. This signal *had* to get through. Right, then, send a general-attention call.

He did, to any Coast Watcher within range—or anyone else on X frequency.

His quick laugh of relief was not lost on the natives. *Kiap* had done it! He had, indeed. A Coast Watcher in eastern New Guinea answered him at once. He had heard Read's frantic calls and was waiting to take a relay. Read gave it to him:

FROM JER: FORTY BOMBERS HEADED YOURS

He had the satisfaction of hearing the unknown Coast Watcher swing into the relay. Once more the whole prearranged machinery of warning went into action. And at Guadalcanal once more the whole deployment tactic found so effective on the seventh—when only a single ship had been hit—went into effect. Once more Macfarlan was the star reporter. Many official records remain to testify to his authentic observing of the most exciting event of his life, for this attack paled that of the previous day in the sheer fanaticism of the attackers and the fury of the defenders.

Again the flight of enemy aircraft was caught between levels of fire, but even more intense fire than previously. The whole arc of the sky was sliced with red tracers and plumed with ack-ack. Three enemy planes at one time plunged smoking toward the sea far below. Another bomber erupted into a solid sheet of flame so huge that Macfarlan was certain that a shell had hit squarely on the warhead of a torpedo in its bay and the whole thing, big enough to kill a ship, had exploded in midair.

Still, through the fury of the concentrated fire, Japanese bombers lived to level off for their torpedo runs—and to launch them. Macfarlan gasped in disbelief, for some of those bombers were flying at maximum speed *between* moving ships, actually below masthead height.

The sea boiled with torpedo wakes. And now it seemed beyond credibility that the defending vessels could live with destruction on the loose against them from every direction. Yet only one torpedo connected: a destroyer was heavily damaged astern and went out of control. Under her own power she dragged herself away. Later, when Macfarlan remembered to look for her, she was no longer to be seen. Now his attention was jerked to another giant explosion. A Japanese warplane had plummeted straight into the deck of the transport. Her whole length quivered with flame, and fire leaped high where the plane had gone through her deck.

Now it was time for the high-up fighters to come in. Once more the smoke of battle and the burning transport obscured Macfarlan's view. But he had seen enough to know that a whole enemy attack fleet had been all but destroyed before his eyes.

Far to the northwestward Read's fingers had teased the essence of excitement from the vernier dial. At about seven megacycles he had come upon strange talk between "Orange Base" and aircraft pilots. Nothing the Australians owned answered to that. The Yanks, then. This was it. He listened again and heard a carrier concerned about getting its fighters all refueled in time to beat off an "expected bomber attack on the transport area."

Read said that this time he really jumped to his feet and cheered. His boys cheered with him. The warning had got through!

He waited, caressing the vernier dials. And soon it came, the intercommunication orders and shouted warnings as the battle closed. The distant voice said that it would go off the air during the battle, but the unseen American was unable to stay off the air and kept breaking in with his own blow-by-blow account of the sight he was witnessing. "Boys, they're shooting them down like flies! I can see one, two, three, four, six—*eight* of them all coming down into the sea together!"

Forty of them had gone over Read's spot. Eight came back. In the two main battles and in at least one side fight later, the enemy had lost more than sixty bombers and Zeros in two days. Read could not know it, of course, nor could anyone else except the Japanese authorities at Rabaul, but the tremendous losses had made impossible the mounting of another sustained attack until new elements had been flown in from the Carolinas some days later.

Despite his crucial role, there was one irony for Read in connection with this victory. While both he and Mason—and other deserving Coast Watchers—would be tendered high American decorations for their work, a British decoration for the "Forty Bombers" message would be accorded not to Read at all, but to an Australian Army officer who happened to be on duty and was concerned with the relay of it from the New Guinea Coast Watcher to Port Moresby and on to Townsville.

In Brisbane we could hardly believe our luck. Conning the maps coupled with an analysis of the reports began to make it dazzlingly plain that Read and Mason were incalculably valuable positions: seemingly every flight that went out of Rabaul droned over Mason's position; all flights originating out of the satellite field at Kavieng went over Read's. The cream of the Japanese naval air power was habitually plotting its own course to its own destruction. If by chance either the two missed them owing to fogs or some temporary deviation in the enemy's navigation, the other Watchers spotted them. It was to be that way throughout the whole Guadalcanal campaign; captured enemy operational data after the war were to tell of the bloodletting there and elsewhere and of the fact that 50 per cent of all the best-trained Japanese air personnel was lost within the first three quarters of 1942.

The solitary unknown Australians out in the islands became legendary figures among the Americans. Both on Guadalcanal and in Brisbane, if you wanted to evoke a smile, it was necessary only to say: "Forty bombers headed yours!" It caught on during the critical months when the enemy would hurl his might in his effort to force the Americans back into the sea, a continuous battle that was to see thousands of casualties on both sides and drain Allied naval strength to a new low before the tide finally was turned late in the year.

The Japanese High Command reacted violently to the events at Guadalcanal. History was to show that a new and much stronger combined headquarters of the Japanese Eleventh Air Fleet and the Eighth Naval Fleet was activated that very day to "prosecute with all vigor the official Southern Operations Plan." As shall be seen presently in a sequence dealing with New Britain and New Guinea during this period, in addition to taking steps to bring in greater power to deal with the Solomons, the enemy already had implemented an alternate to the ill-fated Coral Sea plan to take Port Moresby.

This new threat out of Rabaul was posing multiple headaches for the Allied Intelligence Bureau which certainly could have occupied everyone concerned full time. But we were to be permitted little time to take our eyes off the Solomons. The Japanese wasp's nests there—especially at Lunga—had not been more than normally dangerous as long as they were not stirred up. Lunga's violent displacement by the Allied assault of August 7 produced reactions increasingly dangerous to the Coast Watchers on Guadalcanal—and even farther afield. Calls for help and evacuation began to filter into the AIB decoding center. Japanese remnants in strength were reforming in the coastal areas and in the jungles and they were patrolling in strength. Furthermore, they were venting the anger of their defeat upon the natives. It was clear to their commanders that they had been betrayed by hidden observation agents who probably could not have survived without native permission. Savage reprisals against the natives were adopted to influence them to change their allegiance.

In the northwest part of Guadalcanal Rhoades and Schroeder were in a bad fix. Especially dangerous patrols with cowed native guides had infested the areas of these two. They, in turn, were worried, not only on their own account but about the safety of missionaries and nuns in the vicinity—once they had found the stripped and bayoneted bodies of two missionaries and two nuns on a beach. It was only a matter of time before the native-led patrols would catch up with the rest of them.

In a hasty conference at Brisbane it was decided to shift Hugh Mackenzie from the New Hebrides to Guadalcanal since, in any event, the forthcoming hub of the Solomons war would be there, and effective, mature coordination of AIB operations was requisite to the preservation of our precious people and to serving COM-SOUPAC, which now was bracing for the inevitable counter assault by the Japanese.

Hugh Mackenzie, of the generous mouth and level gray eyes, who eventually would be brought to Brisbane to recover his health and amplify his modest reports, reached Guadalcanal just in time to effect a dramatic—and unauthorized—rescue.

Mackenzie knew that the tough American general of marines had placed great value upon him as a coordinator of Coast Watchers. He strongly suspected that if he made a request to risk a run for it and rescue

Rhoades and Schroeder, he would be refused. Therefore, he did not ask.

"I told Horton that General Vandegrift would skin us alive when he did hear about it," Mackenzie told me later. He had set up AIB's Station "KEN" at the edge of the former Japanese air strip, now called Henderson Field, and he was talking to Lieutenant D. C. Horton of the Australian Navy. Horton was an ordinary chap with an extraordinary way of getting things done. He had asked for the job of rescuing Rhoades and Schroeder. "I'd managed to get a launch," continued Mackenzie. He hitched around and smiled. "Well, the resident commissioner had gone away and ..."

Apparently Horton already had anticipated the requisition *in absentia* and he had already refueled. "It'll soon be dusk, you know," he had said.

They emerged from KEN's fairly snug dugout that had been constructed by them and occupied only a day before the earlier one—a combination of tattered canvas and a hole of warm mud soup—had been blown to bits by enemy marksmanship.

Horton knew the coast; once he had been a colonial district officer. He piloted his little craft to take advantage of every obscure passage and every shadow and at dawn crept into an inlet where Rhoades and Schroeder should have been. They were there—together with thirteen missionaries and a shot-down airman! Everyone turned to in order to camouflage the launch against early-morning air patrols—friend or foe, it would make no difference. During the night Horton repeated his delicate piece of seamanship along the thoroughly hostile coast and landed sixteen grateful people the next dawn.

Later in the morning the telephone rang. "An invitation to call upon General Vandegrift, I'll warrant," Mackenzie said quietly.

It was, emphatically. Horton arose to accompany him. Mackenzie pushed him down. "My party, Dick."

The interview was all he expected. The general had an ample vocabulary, and suffered from no inhibitions. He expressed himself on the subject of discipline and those who lacked it. He quickly confirmed Mackenzie's earlier presumption that operational orders would emanate from no place other than his own headquarters. There were other pointed allusions to this and that and Mackenzie found himself outside where the steaming temperature of Guadalcanal felt gratefully cool to him.

Not long afterward Macfarlan staggered in to KEN from his hide on Gold Ridge. Mackenzie searched his haggard gray features and told him that if he could achieve as sick a look as that, he would put in for home leave—and get it. Obviously it was malaria. How long had Macfarlan had it in such a severe degree?

"About long enough, Hugh. I've clipped my coupons. Brisbane looks good."

He followed Rhoades and Schroeder out by air, his job superbly done, another yet to do in Brisbane. MacArthur awarded him a Distinguished Service Cross. The same award went to Rhoades, and to Read and Mason—all citations marked "Secret." They could not be publicly announced until after the war. Macfarlan was given long recuperation leave, then joined us at AIB headquarters. He was lucky to have gotten out. Soon afterward there were no aircraft to spare. Daily the situation for the American troops grew more critical. Despite extensive losses, the enemy had managed to land reinforcements on the western end of the island and daily the defensive perimeter was being pinched in. A series of violent engagements had tipped the naval strength in favor of the enemy. At one time three American cruisers and one Australian were sunk simultaneously. Carriers went, too. To add to the constant harassment of daytime bombing now came a new threat to life or sanity. Japanese naval units of the Eighth Fleet began shelling Lunga by night. Mackenzie and his little staff were grateful for their stronger dugout, but none had illusions about what a direct hit would do.

Then one night, amid the endless crash of the explosions they had come to know so well, there was a new note. The whole earth rocked to tremendous concussions at regular intervals. Between blasts Mackenzie peered up through the acrid murk and announced: "They've brought in the first team. Those are not less than fourteen-inch guns out there."

He was right. They timed the incoming rounds. They were methodically deadly at twenty to the minute. Nerves went ragged. Heads pounded and seemed to pull apart from the incessant hammering. Once Mackenzie was deep down in another dugout when it was hit. He was flung violently against a wall. He crumbled up. "It was only concussion shock," he said later, "but I knew now there was a limit to what a man could take." Yet even through all of it Station KEN continued to handle the traffic from Read and Mason and the others to the northwest, especially from a hard, dark-miened man named D. C. Kennedy, who was holed up at a plantation on the southern end of New Georgia Island. He was the last one in the chain northwest of Guadalcanal to be able to give warnings; his word was the final prelude to the "red alert" at Lunga.

In Brisbane, Colonel Roberts studied a radio message from Mackenzie. "Kennedy's a bloomin' one-man army, no less," he said. "Look at this."

It was a request from the undefeated, undefeatable man of Segi plantation. He had a loyal following of natives (the natives were worshiping strength here, too, no doubt) and among them all they had slain three barge-loads of enemy soldiers who had ventured too close; captured twenty Japanese pilots who had been shot down in air flights; rescued twenty-two American pilots who similarly had been shot down.

"He wants someone to evacuate his stockade and barracks," chuckled the controller. He picked up a pencil. "Let's see, you Americans estimate that it costs twenty-five thousand dollars to train a pilot. Now then: that's five hundred and fifty thousand Kennedy has saved his gallant allies!"

"How does he rate the Japanese pilots?" someone asked, just as Feldt came in.

"Same difference," explained Feldt. "He has a standing offer to his natives: one case of tinned meat and one bag of rice for every living airman brought in, Japanese or American—only the Japanese are trussed up on poles and carried in that way."

It was an extremely hazardous but well-paying mission for a Catalina a few nights later to effect the evacuation. Kennedy stayed on.

It was the same night that the enemy off the Guadalcanal coast shifted his pattern again, and the two thousand smaller caliber shells that raked the Lunga area were described by Mackenzie as seeming to be even worse than the regular pounding of the big guns. KEN's dugout withstood it, but outside the land was tortured chaos. The next morning they surveyed the damage.

Only one of the three aerials they had so carefully and scientifically constructed for utmost efficiency still stood. But the transmitting wire, in falling, had become entangled in the head of one of the few coconut trees still erect. It looked very much as if KEN would be off the air indefinitely. Still, it was decided to try to raise some comparatively nearby Coast Watcher and arrange for relays of urgent traffic.

The operator warmed up his Teleradio and called. To his astonishment he got a response from one of the alert members of the chain. He asked for a signal strength and readability report. Anxiously they waited. It came.

SEEMS IMPROVED. HAVE YOU BEEN MAKING ADJUSTMENTS?

Mackenzie looked at his men and they looked back at him. Then the strain of the night and those that preceded it broke under their laughter until the tears came.

Between bombings and bombardments they managed to erect two or more receiving aerials. In the first traffic to be snared by them were warnings from Read that the heaviest concentration of aircraft he had yet observed was now on the expanded air strip at Buka village. There were more ships there, too. But why no friendly bombings of them? It was true that the enemy was enjoying some immunity. Southwest Pacific Area bombers were employed day and night trying to prevent enemy reinforcements from reaching the New Guinea coast and the heavy bombers of COMSOUPAC were based too far south to make raids profitable. They could not base closer; there was no fuel for them—and little of anything else except an endless supply of wounded waiting to be airlifted to base hospitals.

Read's reports were followed by grim news from Mason. Literally scores of enemy ships of every kind were anchoring within his range of vision. A tremendous build-up obviously was in progress. Apparently everything now was awaiting the arrival of actual assault-troop convoys from the enemy Caroline Islands.

Then a different kind of message came from Mason. The enemy was out after him and to make sure the Japanese had just landed at Buin crates of savage dogs trained in jungle trailing.

In Brisbane it gave us a shiver of apprehension as nothing else had done. War was always dirty business, but somehow this seemed a new low. Mackenzie pleaded for a bombing mission. AIB urged it. RAAF did it—superbly. Mason's next message included his fervent thanks to all concerned for the beautiful accuracy of a direct hit on the crates.

In a new message to Feldt, Mackenzie expressed fears that this was only the beginning of an enemy attempt to clean out the Coast Watchers. A priority reply went back. Both Read and Mason would get off the air and stay off until ordered to resume. They would hole up in the most secure places inland and avoid capture.

On Maliata Hill, Mason needed no prod; he had it in the form of urgent reports from his native scouts that not fewer than one hundred Japanese constituted the latest punitive patrol that had already set out from Buin to mop up him and his handful of devoted followers.

Mason considered. If the enemy was heavily armed, his progress would be slow in the tortuous country through which Mason planned to lead him. He would bury his Teleradio and travel light. With signs of their occupancy obliterated, Mason released some of his boys to return to their homes for the time being and with the others, he set his face toward the high, mist-shrouded ridges inland. Now his life was in the hands of his boys, whose glistening bodies worked patiently to clear passages just large enough for them to traverse; the jungle closed up after them. It was bitter, exhausting work and that night Mason felt the giddiness of exhaustion. But he was heartened by his scouts who reported that as he had foreseen the enemy had made even less progress. At noon the next day he got it from two observers simultaneously that the Japanese had given up and turned back. He, too, headed back, and by the time the last Japanese had trailed into Buin, Mason's Teleradio was ready for eventualities. He got a permission to transmit if absolutely necessary.

An almost identical experience had befallen Read, except that the enemy out of Buka Passage was more resolute. The second afternoon had found Read's party slogging through a heavy downpour well up into the high interior. They moved with the heaviness of men who were approaching their limit. But there was one more steep ascent before they could reach their objective of a small village on the very top of the highest elevation in the entire area. The miserable trail was almost vertical and slick with black mud and skinned vines. The afternoon was nearly done when they achieved that last wretched lap. At the top they dropped in their tracks to rest on the comparatively open crest.

"It was no choice of mine that we stopped at that one particular spot," Read said later. "We were dead beat and I was on the point of dropping off to nap when I happened to open my eyes for a moment."

The sun had broken through a clearing and a surge of breeze had driven the mist away. Below, in a glorious panorama of a Coast Watcher's dream come true, was a convoy of no less than twelve huge troop transports loaded to the rails with Japanese soldiers and moving toward the southeast.

Read scrambled to his feet.

"Quick, yufelo!" he yelled. "Wairless belong dis felo, you mak um hurry alonga tree!"

His weary boys were galvanized into action. They saw now what the *kiap's* quick eye had taken in. In an ordered chaos of their own they put their backs into the business of breaking out the Teleradio and running an emergency transmitting aerial. For Read it was a repetition of the scene of August 7 above Buka Passage.

From the ridge he counted the ships again. Yes, twelve were in sight. There might have been more, of course. But not one of these was under ten thousand tons. He realized that this convoy was carrying enough assault impact to make bitter business of Guadalcanal. Without a doubt the formidable concentration of warships Mason had been reporting down south was to serve as an advance assault and as an escort for this stuff. He correctly assessed this as the major enemy drive to retake Guadalcanal and all the central Solomons with it.

Everything now depended on the Teleradio. The connections were tight. Read patted the transmitter affectionately. To his grinning blacks he expressed his feelings of how good it was to have the faithful unit. "Wairless i gudfelo samting, dis felo tinkum!" The boys nodded their noisy assent.

The milliampere needle went to maximum and he knew the power was radiating from the jury antenna.

Read had not forgotten Brisbane's order for radio silence. But under the circumstances he was "not very ruddy likely" to observe it. As he prepared to transmit, he said to himself, "Bougainville's a fair size. We'll just keep on the run."

Read's warning reached KEN almost simultaneously with another from Mason that now enemy ships of every description were anchored below him at Buin. Mason listed more than sixty. With this vital intelligence and much more gathered from non-AIB sources, COMSOUPAC fully appreciated that the time of decision had arrived. Nothing could save the worn-down, if now-reinforced, defenders of Guadalcanal unless the naval and air strength of the whole command could be rallied to disrupt the enemy attack before it could hit.

On the night of November 10 the bombardment of Lunga by Japanese warships, doubtless from the concentration Mason had kept under observation, was particularly devastating. At dawn the ships moved back to let the dive bombers take over. That night the ships came in again for one last softening before the kill. But this time they were met off Savo Island by an unexpected American naval force. There was a short, savage fight and the Japanese were beaten off. More American heavy naval units arrived. Repeatedly the opposing forces traded blows. Then one night the main Japanese escort for the transports that Read had seen were met head-on by powerful American naval elements. The enemy was taken under murderous flat trajectory fire at close range. The short battle had no equal for quick, savage ferocity.

This left the enemy transports wide open to attack. It came in successive assault waves from Allied air. The packed convoy was mauled without mercy until it was no convoy—only a remnant that sought to beach itself before total annihilation.

The Japanese reinforcement attempt was a failure. The cost had been dreadful for both sides. But the enemy effort in the Solomons never again would be so great as it had been. Even so, he was still utterly determined, resourceful, and very powerful.

New Year's Eve, 1942 . . .

In Brisbane's brown-out, dense crowds of men and women in the service garbs of half-a-dozen Allied nations jostled and tramped and cheered their way aimlessly along Queen Street, George Street, the City Hall Square, and all the other downtown places of that overpacked garrison Queensland town. Few knew quite why they cheered, for the picture was grim enough. Perhaps it was because they were still alive, or because of Coral Sea, Midway, and Guadalcanal. And there had been another reprieve, too—Port Moresby had been saved.

As has been said, when the Coral Sea victory frustrated the Japanese in their original plan to capture Port Moresby and thus give them an anchor in the north for their encircling movement, the enemy determined upon an overland campaign to do the same thing. For sheer audacity in the face of stupendous natural obstacles, it was typically Japanese. He had landed troops on the north coast of New Guinea at Buna, dug himself in and prepared to assault the Owen Stanley Mountains themselves in order to take Port Moresby from behind. This formidable range dividing New Guinea north and south rose wave after green wave to heights that few white men ever had traversed. Despite concentrated heat, despite disease, despite endless trails that went up sheer, jungle-covered sides and went down sheer sides only to go up again the same way, despite ever-lengthening lines of communication under continuous Allied air attack, and despite constantly mounting opposition in front, this insidious threat had crept to within *forty air miles* of Port Moresby before it dropped in exhaustion. Then it had recoiled upon itself, slowly retreating back over the Kokoda Trail the way it had come.

This threat, too, had originated at Rabaul, that fountainhead of enemy power to the north. And now not only New Britain was increasingly dangerous for any Allied Intelligence Bureau activity, but the New Guinea north coast as well. Plans MacArthur may have cherished for the acquisition of his own bases on the north coast once the Solomons were no longer in danger of Japanese domination would have to await the building of enough force to smash Buna. Then perhaps he could think of a frontal assault on Rabaul. Or, if he could not do that, at least he must provide warning protection for his own flank while he tried to secure such bases on New Guinea as Lae, Salamoa, and Finschafen for his campaign to return northwestward to the Indies and the Philippines— and finally, Japan itself.

AIB had become heavily engaged in New Britain; it was performing a vital piece of work before Buna; it had even recruited, trained, and dispatched its first penetration agents to the Philippines by New Year's Eve of 1942.

But with all this AIB still was tremendously occupied with the critical situation in the Solomons. A very powerful enemy had yet to be contained; he would be slowly decimated—thousands of Allied lives, hundreds and hundreds of riddled aircraft, and whole divisions of Allied warships later....

New Year's Eve, 1942...

Midnight—on a certain small strip of sandy beach on the northeast tip of Bougainville. Absolute silence except for the soft purling of the sea on the sand that showed with pale luminescence under the starlight. Read studied the radium hands of his wrist watch. They merged, straight up.

He raised one arm high, then pumped it up and down. On the beach to the right there was a spark of light. It wavered, nearly died, but ballooned to a bright orange flame. He turned and faced the other way. Already a substantial fire was burning. In back of him twenty-nine silent men and women huddled in the denser shadows of foliage. Read took a long breath: it did things to a man to advertise his presence to the whole world like this after all those months—and especially these last few weeks—of hardly daring to stir without double checking to make certain that a Japanese patrol led by turncoat natives was not onto him. This exposure was a calculated risk of the first magnitude. But there simply was nothing else for it—the civilians must be gotten out of his territory. Without them, he felt that he and his few loyal natives still could carry on; they could continue to radio vital information and throw other barbs into the Japanese who now were stung to true fury by the Coast Watchers who had tipped their hand at every turn in the crucial battle for Guadalcanal. The enemy knew now that he could not retake Guadalcanal; the only thing he could do—and this he would if it expended his last man, ship, and plane—would be to seal it off. In his operational book of "musts," however, was the destruction of the Coast Watchers on Bougainville. Every day the enemy patrols grew more numerous and more savage. There were murder, arson, rape. The natives were fearful and they were turning....

So all through that day there had been a stealthy gathering of civilians summoned by still-loyal natives to this place of rendezvous. Here, if all went well, a gigantic American submarine—the largest in any navy on this side of the world—would appear and take them off. It had rained steadily. The trail had been a horror of mud and slippery leaves and entangling vines. But they had made it—including several nuns of advanced age.

So here the year 1943 was born to the soft sighing of palm fronds and the lisping of the sea.

But wasn't there another sound? From seaward—a faint hail?

Read stared. Then he was sure that he could detect a motionless shadow well out.

"A quid they're hung up on a ruddy reef," he said to himself. From the foliage he pulled out a canoe. Bidding the others stay concealed, he launched the craft and paddled hard. His estimate had been correct. But with good seamanship on the part of the American naval personnel manning the power wherry from the three-thousand-ton U.S.S. *Nautilus* and some guidance from Read, the launch and its rubber raft tow were salvaged from the hidden coral without damage and the beach landing was effected. The submarine herself had to stand well out, for this underseas giant needed more than those dangerous shallows in which to swim safely.

Read directed that the women go first. There were seventeen. One, a nun, whose habit was spattered with the mud of the soaked trail, hesitated, then turned and approached him. She explained that she represented the others in the party when she said that months ago it was considered that he had been needlessly adamant in demanding that they evacuate the island while there still was time, but since then they had learned how right he had been. She apologized for all the trouble they had been. Read swallowed a tight feeling in his throat and muttered his appreciation for her words. Perhaps he had been a bit hard in his direct way; civilians were not such a bad lot, after all.

Then he turned about, unbelieving. At the very last moment one of the women, the widow of a planter, announced in emphatic terms that she had changed her mind and would not go.

"Will you go yourself, Madam?" asked Read icily, "or do you want to be carried aboard, for that I'll well and truly do—*now!*" She went. There was no more room. The launch with its heavily laden tow moved out over the calm sea. Time went by and there was no reappearance of the launch. Anxiously Read scanned the hands of his watch. Dawn was not far off and he knew well that the *Nautilus* would be compelled to clear the area before first light. Soon he and the others would have to put back and get under safe cover. With civilians still to be looked after it would be grim. Of course there were other civilians on the island, but they were not in his immediate territory; things were just too hot here. Finally, the launch returned and just before dawn broke Read was able to send a message from the Teleradio hidden on the beach: "Mission accomplished."

He and his natives turned back into the jungle, happily laden with chocolate bars, cigarettes, tinned hams, razors, mirrors, tooth paste, and a variety of other things supplied them by the generous gobs of the *Nautilus*. Even the conviction of enemy reprisals to come failed to dim their New Year's party in the bush.

It was a short respite. The enemy was implacable. His patrols became constant. Nor was he afraid to come inland. The key to his efficiency lay in his impressment of natives, now demoralized by a combination of propaganda that was superior to the Allied brand, and brutality if there was a lack of cooperation.

There was a flurry of AIB priority messages involving Mason. From Guadalcanal, Mackenzie told Brisbane that reliable information indicated the presence of not less than four enemy patrols of some fifty men each converging on Mason's priceless hide above the Buin anchorage. There was no time to lose. Mason was ordered to bury his Teleradio and anything else he could dispense with and go light but go fast. He would try to effect a junction with Read, a hundred and thirty miles of the most rugged walking through the bush—to say nothing of the enemy and his hoard of native aides. It was learned later that Mason had barely cleared the old site when the Japanese swept down on it from all sides; he never would have had a chance.

Read knew Mason only through the medium of the Teleradio. But the combination of electronics and his own intuition added up to what he felt was a solid enough appraisal. Somehow, he was sure, Mason would get through. Anxious weeks later he reported that the jungle telegraph was active, and that he now was able to trace the progress the little safari was making along villainous trails made doubly hazardous by repeated traps placed by the enemy. Read decided to send help.

From the beginning, as the reader will recall, there had been on Bougainville a detachment of twenty-five Australian soldiers who had been assigned to guard supplies accumulated there prior to the outbreak of war. Read queried their officer and got agreement that they would work southward to try to contact Mason. It proved to be sound tactics, for in due time he had word that the junction had been effected somewhere in the wild bush and that the combined party now was proceeding north.

Then came a message from the soldiers that left Read disturbed: for reasons of his own, Mason had sent the main party forward while he and one police boy detoured to investigate a camp of refugee Chinese. If that message bothered him, the next one produced genuine anxiety. It said that shortly after the separation had been effected, a strong enemy patrol had been spotted between them and Mason and now he was cut off. The soldiers wanted instructions. Read considered all angles and then, examining again his preconceived assessment of Mason's character sent: "Wait."

For the next three days Read's boys squatted on their haunches and watched their *kiap*. His dark-gray eyes stared unseeing into the jungle and his jaw, with a strong hint of stubbornness in it, seemed to set even more firmly. The boys looked at each other and there was admiration in their glance. Here was a *kiap* worth sticking by: he was one to hold fast when action could be foolish.

And then came the runner. The party was just below and coming. A *kiap* was with them, limping badly.

Read and Mason met, and Read knew how accurate his appraisal had been. The signal he had been waiting to send to AIB was transmitted by Read just six weeks after Mason's own Teleradio had gone silent at Buin.

One of Mason's feet was in bad shape. When he had become separated, he had taken refuge with a white miner who had been successfully evading the Japanese since the beginning. A foot blister had become infected and was suppurating. The miner tended him for two days, then Mason had hobbled off. Bougainville had beaten strong men down to defeat despite every assistance, yet this planter who boasted none of the physical qualities of the campaigner now proposed to make the last lap of it on only one good foot.

He did it. Read and Mason measured each other and liked what they saw. They discussed the situation. They were in immediate agreement on two points—as many of the remaining civilians, including refugee

Chinese, should be evacuated at the earliest opportunity, and the soldiers should also go, since they were in poor shape owing to prolonged exposure to a kind of rugged tropical life to which they were not acclimatized. They "made a signal" to Mackenzie, who concurred and asked Brisbane's okay. Brisbane sent it, but said that the soldiers were to be replaced by fresh men from New Guinea who could give the Coast Watchers some protection. In addition, there would be three new officer Coast Watchers. In that case would Read and Mason come out for a most thoroughly earned rest?

The two men looked at each other over mugs of tea. Outside the hut of bamboo stalks and fiber roof the rain drummed down in a gray blur. Read shivered slightly and looked into his tea mug, then at his faithful police boys in their jungle-stained khaki shorts. Their eyes were on him. He looked at Paul Mason, who smiled slowly and shook his head. Read put his cup down and chuckled. Obviously they agreed that neither they nor their black boys would take to the life of a "base wallah."

He flipped the Teleradio to "on," and Brisbane got its answer.

The next evacuation was set for the last of March at the same place on the northeast coast. Then came the word that this time it would not be the *Nautilus*, but an operational submarine with compressed spaces that certainly were not designed for the accommodation of passengers. That drew another signal from Read. It was one that jolted Brisbane. He had no less than twelve women and twenty-seven children in addition to twelve soldiers in the first contingent for exchange.

The signal that came back from Brisbane gave no hint of the soul searching that had gone into its making: the soldiers would have to wait. Read sighed. So be it. Some of the civilians had made the trek of a hundred miles to the rendezvous on foot.

The next day Read supervised the concentration of his charges at the same spot as had sheltered the group on New Year's Eve. Quietly he spoke to the commander of the downcast soldiers, telling him to have the men ready "with their kits." The other looked at him sharply. Read shrugged. He muttered something about man's inability to forecast the future accurately if at all, and said good-by for the moment.

What happened that afternoon was not what Read had anticipated. A short time after his conversation with the troop commander he was staring at the small Japanese ship that had come into the very eye of the evacuation area and dropped anchor a few hundred yards offshore. What a situation! Here he had gathered all the weak and the helpless he could sweep up and delivered them straight into the hands of the enemy. And the submarine as well.

He was everywhere at once. There were men, women, and children to hide in the jungle; there was a small fleet of canoes to conceal. There was a warning message to be sent to AIB at Guadalcanal to warn the submarine not to come in. And if sheer mental persuasion had power, then Read bent every iota of his to influence the Japanese to stay aboard their ship and leave without landing.

The day wore on, and with each measure of light that went out of the sky Read's hopes rose another notch; surely the enemy would not land in the darkness. But why were the Japanese here at all? Were they going to establish a base, or rendezvous with another ship?

There was no sleep for him that night.

Shortly after dawn the enemy ship hauled in her anchor chain, her crew sluicing off the links as they came inboard. The vessel slowly gathered way and left. He sent another message. It would be tonight. Meanwhile, they could wait—and rest.

At dusk the signal fires were lighted. Almost instantly the submarine rose out of the depths and shook the sea from her. Read launched a canoe and went out.

Lieutenant Commander Foley of the U.S.S. *Gato* was chuckling. The *Gato* had spotted the unwelcome visitor immediately. "He can thank his lucky stars you and your party were there hiding on the beach," he had said. "Otherwise we would have blasted him sky high. We were practically sitting under him all the time!"

Read cleared his throat. "I..." he began, looking straight at Foley, "have fifty-one persons to go out." This figure, of course, included the weary soldiers.

The American's eyebrows went up. Then he grinned, called his executive officer, and discussed it with him. He turned back.

"If we can dog down the hatch over the head of the last one, we might say we're full. Bring 'em on!"

There was a story that the last six were made comfortable in the torpedo tubes.

Now on Bougainville there remained the last of the soldiers and a Bishop Wade and some of his people. Read sent word for them to be ready for the next and last evacuation party.

It was in the interval between the second and the last evacuation that Read observed nearly one hundred Japanese Zero planes take off from the greatly expanded air strip at Buka village and strike off for the southeast.

Again his famous signals preceded the raiders. The ensuing sky battle off Guadalcanal was one of the most devastating the enemy had yet experienced, for there was no doubt that he already had expended the cream of his experienced pilots while the Allied fighters were just achieving their maximum efficiency, and with their added advantage of ample time and height, the issue was certain.

The last evacuation went off without a hitch. Now no civilians remained except some refugee Chinese and possibly a few white men fairly well able to take care of themselves. A new burden had been laid upon Read, however, for a RAAF Catalina in attempting a resupply drop near the high country had crashed into a mountain. Some crew members had been killed outright. A brave but futile attempt had been made to carry the critically injured survivors to the submarine evacuation spot before the ship left. A camp inland was established for them. On the other side of the scales there now were several fresh, experienced Coast Watchers. They included an old New Guinea friend of Read, one "Wobbie Wobinson"—Captain E. D. Robinson of the Australian Army was unable to articulate the letter "r"; result: "Wobbie."

Stories of "Wobbie" were legion. Feldt later told of the purported initial conversation between him and Read. It had gone like this:

Read: "I had it first that the 'Robbie' coming in was 'Dry Robbie.'" He was referring to the only non-drinking Coast Watcher in the whole organization, Lieutenant H. A. F. Robertson of the Australian Army.

Wobbie, emphatically: "Not I, mate. I likes me pint." (In peacetime he had owned a pub in Sydney.) Then he asked Read if he had heard how convivial indulgence in the cup that cheers had done no less than save the very life of still another "Robbie." This time he meant Flying Officer R. A. Robinson on Mackenzie's staff at Lunga. Read had not, and awaited elucidation.

Wobbie: "Well, now. Y' see a wuddy Nip ai'cwaft that had been plaguin' th' Yanks had just come back t' pay anothah visit and th' Yanks was lettin' him have what-foah with all the ack-ack in the place. The wed wa'ning signal had been displayed all th' time but ouah man was havin' none of it because th' beeah was on that day. So he was sittin' theah and he had just tilted his head back t' get the last fine twickle o' it, when a wuddy hunk o' shwapnel comes in and actually nicks his wuddy fo'head. Now then, it's plain t' be seen that if he'd been sittin' in a nondwinkinf position—like 'Dwy Wobbie' would, n' doubt—he'd now be flyin' about by means of wuddy wings of his own.... Let that be a lesson t' you, Jack Wead!"

Other things also had been occurring at Lunga. The original Marines had been relieved during December 1952 and had joyfully pushed off for Australia. With them had gone Major General A. Vandegrift. Major General A. M. Patch had replaced that warm friend of the Coast Watchers as commander of ground forces, now predominantly army infantry. To Mackenzie's relief, General Patch was no less solid. The same could not be reported of his intelligence officer. Totally untrained for his job, this individual was described by Mackenzie as one who "did not seem able to absorb much knowledge of the geography of the Solomon Islands. His plan was based on his desire to rectify the unsymmetrical appearance on his map of the colored flags indicating Coast Watching stations." The G2 put his "plan" before General Patch. It is not known whether it was at this time that the American commander sent a message to Washington requesting the assignment of the most experienced Intelligence officer they could lay hands on (he got him in the person of Colonel E. E. Brown) but it is known that he sent for Mackenzie and assured him that no one—and he meant no one—would disturb the Coast Watchers except Mackenzie himself, and that thereafter he should consider himself as having direct access to the general's command post at any time.

It had been a gratifying moment. But Mackenzie found himself increasingly unable to rally to good luck when it did occur. The frightful ordeal of November and the unceasing strain, coupled with low-grade malaria, were taking their toll. At last he radioed Feldt. Feldt decided to investigate in person. He flew out. A few days later we were dismayed at Mackenzie's urgent message of critical illness—not of himself but of Eric Feldt. The untiring chief of the Northeast Area section had collapsed while flying in a small military plane. He now was in capable American medical hands. Diagnosis: coronary thrombosis. Prognosis: fair, under conditions of absolute rest.

Eventually he was flown back to Brisbane and hospitalized. It was the end of active military service for him. Yet from his sickroom he eventually would compile the finest complete story of the Coast Watchers to come out of the war.

The casualties of war were not restricted to the fighting fronts. In Brisbane devoted, dogged Colonel Merle-Smith also had passed the limit of physical elasticity. He was told to rest, and while reluctantly complying, suffered a heart seizure. Flown back to mainland United States, he died in New York.

On Guadalcanal, Mackenzie persisted through a haze of exhaustion until Commander I. D. J. Pryce Jones could take over at KEN. Then he was evacuated to Australia. The drain on his vitality had been too much and the low-grade malaria degenerated into feared "blackwater" fever. He was critically ill for months. Upon his eventual recovery he came to us at AIB headquarters in Heindorff House, Brisbane, where his quiet friendliness and unassuming efficiency would be strengthening factors at all times.

Meanwhile a slight man with snow-white hair, frosty eyebrows over friendly pale blue eyes, and a granite-like jaw had taken over from Feldt. Commander J. C. McManus was not an "islander" but he was as capable as he was fair, and he was a champion of everything and everyone related to Coast Watching. It was plain that he would be accepted, even by the clannish, unpredictable, independent "islanders" themselves.

A development that was to have far-reaching effect on Coast Watching had its inception at the time he took over. This was a plan for greatly expanding the scheme which Kennedy on Segi Island had been employing so successfully against the enemy—the recruitment of volunteers among qualified natives to serve as trained, officially enrolled guerrillas under the command of the Coast Watchers themselves. Eventually the plan would result in the development of splendid battalions that would become the terrors of the Japanese, on New Britain and New Guinea as well as the Solomons; they would account for the bulk of the more than fifty-four hundred enemy killed and nearly fifteen hundred wounded officially credited to the Northeast Area section. The high proportion of killed to wounded was significant.

It fell on McManus, also, to implement an earlier decision to revitalize Coast Watching on Bougainville. One of the newcomer Coast Watchers was Lieutenant J. H. Keenan of the Australian Navy. He was to take over Read's old spot in the north, while Read and Robinson, together with some soldiers in support, were to locate on the east coast in a more central area. Mason would head south for his old location, which seemed to be quieter now.

Any appearance of peacefulness was wholly deceptive. Hardly had Read gotten into position and Mason traversed a part of his rugged way back when trouble was reported from every point. Some of it was because of the inexperience of some of the newcomers, but mainly it was the aggressive enemy and his host of cooperating natives. Mason, hemmed in and having lost one of the new officers to a Japanese bullet, turned in his tracks and set his course for a reunion with Read or the Buka force. But the latter had been attacked, too, and Lieutenant Keenan had escaped only by prompt evasive action. He later returned to his burned-out site and prepared to continue reporting movements at Buka. But the enemy in turn was watching him.

Then the lightning hit Read.

The observation camp he and Robinson had set up was high on a ridge. Some miles to the westward were the soldiers. Japanese were known to have been in the area and Reed, the old hand, employed every device of his jungle wile to prevent surprise. An "outer guard" was established at some distance from his lookout along the only trail that would lead directly into it. Then came a stretch along the trail cunningly laid with dry bamboo which would crack with a discernible report when inadvertently stepped on. At the end of this lane and covering it with fire was the "inner guard." Then the camp itself.

All had been quiet enough. Except for a visit by an aged native who proffered taro in exchange for some calico, nothing unusual had been noted. Thoughtfully Read watched the old native depart. He glanced about the shack with its palm-thatch walls. Almost unconsciously he gathered together a few bits and pieces and stuffed them in his pack.

Shortly thereafter Wobbie reported that a runner had come in "with a yarn about a strong party of Nips coming this way." After hasty conference it was decided to bury the radio and otherwise prepare for hasty departure should that be indicated.

It was done, but not before runners had been sent to warn the soldiers farther west. As dusk came down, the guard was set up, a stronger "inner guard" than usual being maintained. Faithful Sergeant Yauwika was in charge. With the darkness came the multiple sounds of the jungle and the unseen tautness of danger.

Suddenly there was a single shot from the outer guard, followed almost at once by the firing of automatic weapons. That would be the inner guard. Then came explosions from half the circumference around. Grenades.

Action, right action, was the rule of survival for the successful Coast Watcher, and Read had been successful. His instant estimate was that it would be quick death either to remain in the hut or exit from it through the doorway. He dived headlong through the side of the hut itself on the side leading down from the ridge. He landed rolling down the declivity, and after him came Robinson. He managed to check his momentum and Robinson brought up against him.

Now they could hear the explosions of grenades being hurled into the hut itself. Read wondered what had happened to Yauwika. The firing seemed to be general except from below the slope. Accordingly, they began to make their way down. From the glow of orange light above them they knew that the enemy had set fire to the camp. Read thought: *This time they've got us.*

The declivity grew steeper and they were sliding, feet first. Read grasped the undermatting of the jungle and checked himself. He tried to dig in with his boots, but there was no earth beneath them.

"Wobbie, for God's sake don't come any farther," he whispered.

There was nothing to do but hang on.

It was the longest night Read had ever known. But they were alive. And then after an age of weariness, of biting mosquitoes and crawling things, the night ended. Read could see that he was hanging on the eyebrow of a true cliff.

Slowly, cautiously, he pulled himself up, hardly daring to put an extra pound of pressure on the rotted matting beneath him. Inch by inch he gained and together he and Robinson pushed upward until they could rest.

It was useless to think of climbing back to the camp; the enemy would be waiting for just that move. They decided to work to the left and try to get down. Eventually they found a way, and for two days they rested on the floor of the valley. Then they went on. Read was grateful for his own foresight in caching food where he knew he could find it. Later they cautiously revisited the burned-out camp. Yauwika had escaped into the jungle, and they were reunited. To their joy the faithful Teleradio also was recovered unscathed. But the batteries had been found and smashed. They were about to leave when Read stopped short and stared. Before him, at the edge of the lane of fire of the former inner guard post, was the body of the old native who had visited them that fateful afternoon to trade taro for calico. His stealthy tread on the dry bamboo as he guided the enemy in had alerted the sentries. ...

They slung the Teleradio on poles and carried it in a search for the troops, in time finding them. What's more, they dug up a new set of batteries and a charging engine. But there was no gasoline. A few days later Keenan joined. Read looked beyond him. There were a few native scouts and police boys, but no Teleradio.

"It's buried," said Keenan. "The blighters hit me again and there was no time to do anything else."

Read could sympathize with that. "Petrol?" he asked.

"Might be a bit in the ground."

Read considered. He knew as well as he knew anything that Coast Watching on Bougainville was finished. It was not a question of staying useful any longer—but of staying alive. He must get gasoline, charge the batteries, and signal Brisbane. It would be a signal that he hated to send— a request for evacuation for everyone who still was left before it was too late. He did not know that it was already too late for some. A force of about eighty Japanese had wiped out one whole element of the soldiers together with the injured RAAF men in the west. But he did learn that Mason was unreported. This time he could not feel the old faith in Mason's indestructibility. To elude for the second time an enemy net covering the frightful country Mason would have to traverse to rejoin them was "stretching it a bit thin, just a jolly bit thin."

Read sent Keenan and his party westward with instructions to hunt for one of the new Coast Watchers named McPhee and any of the surviving soldiers; they were to gather at a point previously agreed upon with the United States Navy as a pickup site should a submarine be available. Read well knew that the submarine would call if only he could charge his batteries long enough to make contact with AIB at Lunga and request it.

Sure enough, there was a quart of gasoline at Keenan's old camp above Buka. Read transmitted KEN's call and felt a sweat of weakness come out on him when Lunga answered. His message began: "My duty to report that position all here vitally serious . . ." It ended: "Reluctantly urge immediate evacuation." AIB's concurrence was instantaneous. Lunga was directed to request COMSOUPAC for a quick submarine evacuation. COMSOUPAC needed no prod; their collective estimate of the Coast Watchers was unconditionally the highest and submarine crews considered it a privilege to serve the men who had served everyone else so long.

Now began the deeply secret move westward to effect the rendezvous. It would be a most dangerous undertaking, for not only were there Coast Watchers but soldiers, and even the last of the Chinese who had been hiding from the merciless persecutions of the enemy and of the now almost completely subverted natives. In addition, Read had stood solid on another point: Every loyal native carrier and police boy "who wants to should go with us when we leave."

His batteries were still good for a few more schedules, and he was grateful indeed that they were, for it was while they still were trekking across to the westward that he heard that the unbelievable Paul Mason had made it again after a nightmare of living off yams and pawpaw, of endless wild trails and enemy traps. The same message concluded: "Submarine can be made available on four days' notice."

Read sat down wearily and encoded the reply that he knew would end his Coast Watching in Bougainville within four days. The faithful radio transmitted the instruction for Mason to take complete command and embark everyone at the earliest possible time while he and his own party would make for the coast and a lift whenever the submarine could return for them. Then he called his boys around him and explained the situation. There was much quiet talk among them in their own dialect and then all turned to and broke camp for the last time. Read was wise enough not to question. But he knew that some of the married police boys would not be among those present on the beach when the submarine sent her rubber rafts ashore.

Read was still on the trail when Mason's big party of soldiers, Coast Watchers, Chinese, loyal natives, and two remaining survivors of the crashed aircraft assembled on the appointed beach. Mason sensed the tension. He shared it. For somehow, after seventeen months of blandly defying fate, it seemed to be stretching it too fine to expect that now, crowded as they were into the last few precious hours and the last few fugitive miles, they still could play it out and win.

How the U.S.S. *Guardfish* stuffed all sixty people in her narrow steel gullet and put to sea no one quite knew. She arranged a transfer to a small surface naval craft destined for Guadalcanal and turned about to keep an appointment farther north with Read. It would take a little doing, for this time she was to go in over uncharted reefs. She did, and paused offshore with her snout pointing fair between the two points of yellow light that suddenly had bloomed where the dark mass of the land met the sea.

The *Guardfish* took her passengers aboard and put about. In the control room Read and "Wobbie" blinked to accustom their eyes to the glaring white interior. The hatch snapped down. Jack Read was not seeing the clustered trunks that carried the nerve system of the ship, nor the bland faces of the twin-depth gauges, nor the "Christmas Tree" panel on which the red lights were blinking to green, for it had come to him that the hatch in snapping down like that had snapped shut on a whole chapter of irregular war in the Pacific. He sagged inwardly. He had done his best, and dozens of other Coast Watchers had done theirs, but God was on the side of the strongest battalions. What he could not see then was that strength itself was shifting. He was too tired, too drained from having done his part to deny Guadalcanal to the enemy, to even imagine that within months the reinforced Allies would smash northwest into the New Georgia chain, and then, finally, Bougainville itself, and that when this was done, still another chapter of irregular warfare would be written. This time, for other and later Coast Watchers, there would be no enforced last-minute escapes from the last few yards of safe ground, for this time, as in mid-1943 when Read left Bougainville, God was *still* on the side of the strongest battalions. He would learn of AIB's new penetrations though, for after a brief recuperation he would go to New Guinea. In the dried-grass huts clinging to the bare, sun-baked hills where the powerful AIB radios were hidden he would hear all about them.

He would hear of "Wobbie," back in the fight, arranging to get his beloved beer—from a Japanese ship on the beach. Of youthful Henry Josselyn, who knew from the flashes on the clouds that the battle-wagons were slugging it out and that there might be survivors, so he sent out his boys to look; they came back with one hundred sixty-one survivors of the sunken American cruiser *Helena*. Like Kennedy, Josselyn took in airmen as well and got credit for thirty-one Americans and twenty-two Japanese. As for Kennedy, he stuck it to the last, retreating only when too many United States Marines were sent in to protect his position!

Read learned about the superb work done by that explosive Irishman, J. A. Corrigan, who got an American Legion of Merit for his actions. He learned, too, that Keenan went back in, as did "Snowy" Rhoades. The latter guided in American assault infantry and then became one of them to even up a few scores. There were more, many more.

He would, of course, also hear of the things that did not go so well, for while the telling of it this long afterward minimizes the troubles, the examples of carelessness, of indifference, or poor direction, and worse coordination, and all those other manifestations of the perverse and the crude in the make-up of the human animal— these were there then as in any other war in history. There was, for instance, the necessity for pulling out Lieutenant A. R. Evans of the Australian Navy from his Coast Watcher post on one of the central Solomon Islands owing to what appeared to be the total disregard of his information by American authorities who needed it most and the persistent American bombings of his position and that of the natives who were helping him over the whole period of his occupancy.

It remained, however, for neither enemy action nor our own blunders to account for an outstanding failure of a Coast Watching station in the

Solomons to accomplish its mission. The trouble was attributable to that allegedly deadliest of the species, the female.

In his official report on the case, Mackenzie was characteristically gentle. Said he: "To what extent the attractions and generosity of the Polynesian women [of a certain island], who greatly outnumber the men, contributed to the general slackness of the Coast Watching station there is a matter for conjecture." He added: "The number of aircraft which developed engine trouble when passing [the island] and were 'compelled' to alight on its hospitable lagoon eventually caused a certain amount of official concern."

Yet despite all the difficulties, the enemy was being slowly forced back northwestward.

This development improved our position in the Solomons and the lot of our Coast Watchers there. But during the first phase of the war, that concerned with the enemy's attempt to seal off Australia, AIB had also infiltrated agents into New Britain and New Guinea. As the enemy was displaced in the south, the position of our Watchers in these northwesterly areas became more hazardous. In order to tell the full story of AIB operations in New Britain and New Guinea it is necessary to turn back now to 1942.

Native Privates Simet and Marcus, among Royal Australian Naval Volunteer Reserve wireless personnel at the Allied Intelligence Bureau, Wunung Plantation, New Britain.

Tamkaidan, New Britain. 1945-01-15. Lieutenant J. Sampson and a native member of the unit showing guerrilla equipment issued to members of the Allied Intelligence Bureau. Seen are: rifle, pistol, ammunition for both, compass, grenades, water bottle, knife, field dressing and medical outfit. In the background is a captured Japanese light machine gun, Type 96, which the unit is using against the enemy.

Part 3

NEW BRITAIN AND NEW GUINEA

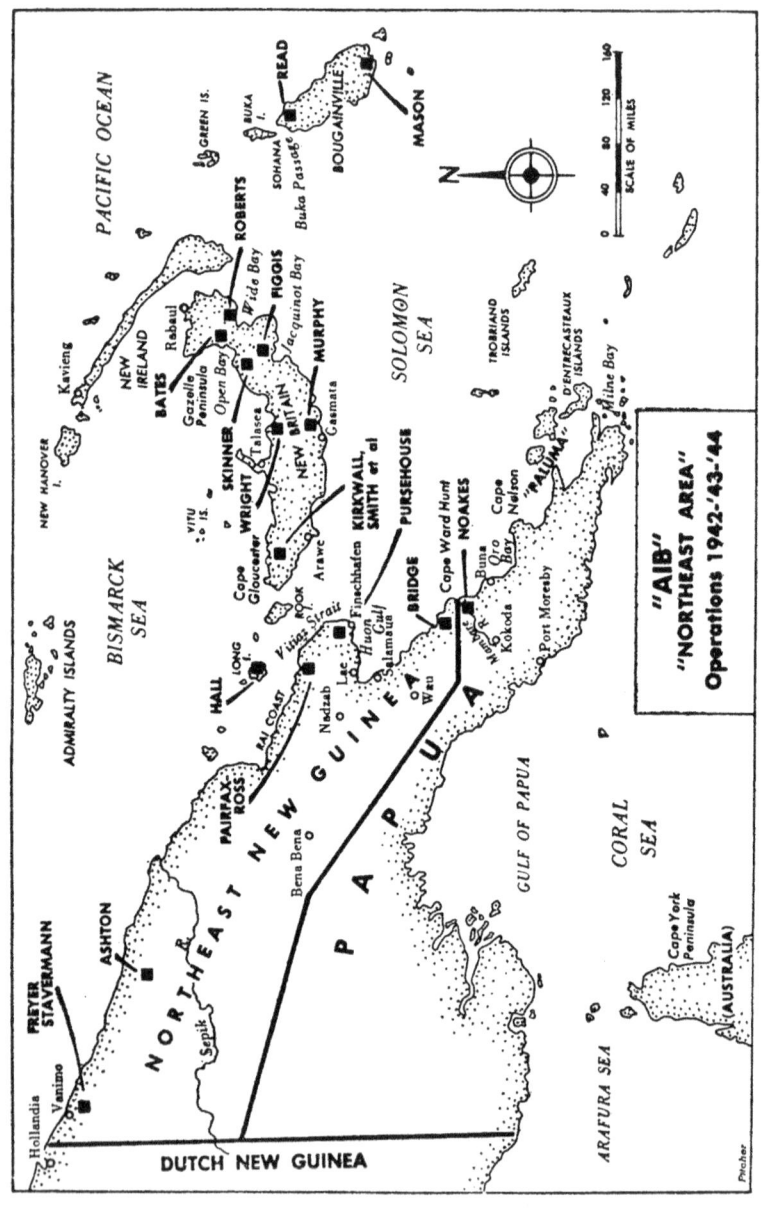

New Britain Interlude

OF THE MANY STARK REALITIES FACING GHQ SWPA in mid-1942, none was more painful than the enemy's firm possession and strong military build-up of Rabaul, at the northeastern tip of New Britain. The wide harbor of Rabaul actually was the sea-filled crater of a huge volcano that long ago had blown out all but a portion of its cone in what must have been an enormous eruption. The town itself once had been the center of pre-World War I German colonial administration. Mandated to Australia after the war, the island of New Britain—and New Ireland beyond it to the northeast—interested the tough Australian "islanders." Rabaul, located on the north and eastern sea-front flats from which the land rose to the remnants of the old volcanic cone, grew modestly under their ministrations. There also was a Chinatown with a population of between five and seven hundred. There were perhaps twice as many whites in the area and several times more Melanesians of various degrees of civilization. Just before the Japanese war broke out, the remnants of the old volcano erupted, not once but twice. In November of 1941, when our survey mission from Manila visited there, the air was still chokingly heavy with pumice dust.

There was a squadron of obsolete Australian fighter aircraft there, a battalion of troops, and two six-inch harbor defense guns.

Life for the military was unenviable. There was but one compensation, the senior officer explained: they were three thousand miles from headquarters at Melbourne and a man did not have to live in fear of the telephone.

It was one of those ironic coincidences that he hardly had uttered the words when a courier aircraft was descried approaching from the south. The weary flier was trundled in by a small truck white with pumice dust. He saluted smartly and presented a heavy Manila envelope, the heavier for the red seals impressively fortifying the inscription that he was "On His Majesty's Service." The commander excused himself, obviously in some nervous anticipation of what might be justifying such urgent dispatch over land and sea and lan^ again. As he read, his already sunburned face turned apoplectic. Wordlessly, he indicated that we were to read. As nearly as I can recall the words, they were:

> Immediately upon receipt of this you will proceed with the construction of six pan latrines according to the enclosed specifications and when completed you will report in detail... etc....

It could happen in any army!

Unfortunately for him and the rest of Rabaul, the messages the Japanese were to drop only a month later not only silenced any possibility of effective reply to them, but were the forerunners of many others that would blast the tiny garrison out of the place and send its survivors scattering down the length of New Britain in a bitter retreat toward the possible safety of New Guinea. Among those who were compelled to go were many who ultimately would live to fight back as Coast Watchers, such as Hugh Mackenzie, Macfarlan, a stocky, rock-faced man called "Blue" Harris, and a dark-haired chap named Peter Figgis. Meanwhile the Japanese swiftly consolidated their gains.

As has been said, it was obvious to GHQ from the first that as long as the bastion of Rabaul, backed by Kavieng, was permitted to exist as a Japanese Gibraltar for directing and supporting the Japanese encircling operations, it was useless to think of attempting to establish Allied bases on the north coast of New Guinea for fighting back to the Indies and the Philippines without first providing continuous flank protection. Even more to the point was the fact that in all probability such a strong hub of operations could keep GHQ dancing to the shrill minors of oriental martial music at any time the Japanese High Command elected to pipe it.

The reduction of Rabaul, the heart from which enemy strength was feeding into the Solomons and could feed into New Guinea, appeared to be a priority project.

But before an effective assault could even be soundly planned, much more would have to be known about the enemy dispositions there: his strength, his defensive capability, and so on. The old problem of intelligence information.

Could AIB help by sending in observers, who not only might supply such information, but remain to act as Coast Watchers to protect New Guinea, just as those in the Solomons were checking movements out of Rabaul in that direction?

It all depended upon the attitude of the New Britain natives.

And so it was that in mid-1942 Lieutenant Malcolm H. Wright of the Australian Navy became a human guinea pig to test the natives' reaction to his presence. He would be inserted by a United States submarine and thus would become the first of many GHQ agents to utilize this modern version of ancient Troy's Trojan horse for the accomplishment of espionage and sabotage.

There was something of perpetual youth in Malcolm Wright. Dark, merry-eyed, he radiated a certain basic friendliness that was quite irresistible. Like Read, he had been a field employee of the colonial government, originally in New Guinea. Read was assigned to the Solomons and Wright quit his job in New Guinea to join the Royal Australian Naval Reserve and was taking anti-submarine training on the mainland when the Japanese opened the war. He was so keen to be sent back to New Guinea that he deliberately failed his examinations to make sure he would be available for assignment. At that point he joined "Ferdinand" at Townsville, on the northeastern coast of Queensland, Australia.

"Ferdinand," let it be said, had become Feldt's own designation for his Coast Watcher organization. Quite shamelessly he appropriated the symbol of the benign old toro with flowers entwined in his horns as the motif of his "escutcheon" for Ferdinand. (I still preserve the original drawing.) He reminded his recruits that like Ferdinand a Coast Watcher never was to fight unless flushed out of his flowery fields and compelled to do battle in order to save his life; otherwise he was just to sit quietly and look.

Wright pleaded for the Rabaul assignment and got it. One of the oldest of the hard-pressed United States Navy's submarines, the S-42, awaited him in the Brisbane River. He turned for a farewell look at the familiar skyline of Brisbane, then indistinct in brownout. Malcolm Wright was a normal man and suddenly he was experiencing the pull of the herding instinct, the desire for security with his fellow men, the ways of peace and a quiet quaff in the pub of an evening with friends. He dropped down the conning-tower ladder, the imprint of that skyline still in his mind's eye.

Then one of the gobs reached for his meager gear to help him stow it. He grinned and the gob grinned back. And suddenly it was all right again. People liked Wright.

Deliberately he set out to rub the image of the Brisbane skyline from his mind and in its place to create one of a tropical beach, of slender palms leaning out over the waters of Adler Bay, only forty miles from the center of all Japanese invasion activity for the south and southeast. There should be a few native huts there, but there should be something else quite incongruous in that tropical scene. He could not be sure how many Chinese there would be—a dozen, two dozen? As he lay in his tiny bunk that night while the old S-boat thudded to the labor of her over-age Diesels, he realized that the clammy perspiration all over his body was not owing solely to the closely confined space of the submarine nor the fact that she was traversing the near-equatorial waters of the Coral Sea, but rather that he was experiencing his first case of "having the wind up." He was plain scared. He declared that it had come from reaction after all the excitement incidental to his "dispatch" (there were expressions of broad humor around AIB concerning the use of this operational term with its implication of finality and in some areas it was discontinued.) His normally buoyant spirits rose to support him again. It was true that he had no sound estimate of the feelings of those Chinese who had been forced to flee Rabaul and leave their businesses, shops, comfortably stuffed warehouses, and their homes to the invader, but it was safe to assume that they were not happy with the Australian Government about it. Nevertheless, it still would be his job to contact them and try to convince them that their best chance of getting any of it back lay in cooperating with him, giving him every bit of information they could think of about Rabaul and the military situation generally in north New Britain. Doubtless they would be armed.

But why think of it? He started to turn on his side and struck his head on the hard steel surface of his bunk mate—a bulbous torpedo. Intended for coastal service only, these little S-boats were among the oldest in the Navy and they were the last word in discomfort for their crews—cramped, reeking with mechanical and chemical fumes, and veritable sweatboxes without air conditioning. Yet they had made the long voyage across the Pacific, and every now and then these relatively feeble Davids sent a Japanese Goliath to its watery grave.

Running on the surface at night and beneath it in the daytime, the submarine at last made its way into Adler Bay. Wright had fitted in well with the submarine crew—a good test. He had learned to relax, too, but now as he fixed his eyes to the periscope for a sub-surface survey of the beach, his face became serious. Spray drenched the hooded lens and the beach itself was blurred. He could make out little and he could see no signs of an encampment. There appeared to be, however, only a moderate sea running, although he and the skipper had gone all over it many times and he knew there would be tricky currents.

The skipper had eyed Wright's equipment: an Australian .303 military rifle, a flashlight, canteen, a week's tucker, a mosquito net and ground sheet, wax matches, sheath knife and shaving gear, and had remarked that it seemed rather meager "for a week's sojourn in the country."

Wright had quipped back that it was due "to the bloomin' rationing, mate."

When it was fully dark, the S-42 surfaced. Wright's small landing craft was readied and crew members held it steady against the possibility of damage against the steel sides of the submarine. There was a repetition of the time and place for the return rendezvous, and with muted expressions of good luck the Americans shoved him away.

Immediately wind and currents took charge of his frail craft. Wright gripped his paddle and put his back to it. But to his amazement he discovered that he could barely make seaway. He would have sworn that the weather was not unruly; Malcolm Wright was to learn much about the waters off New Britain before the war was over.

It occurred to him that he might spin in a circle and ram the submarine. But the S-boat had vanished as completely as if she had never existed. There was a low, driving mist that belied what shortly before had looked like a reasonably clear night. With every rise he twisted this way and that to get his bearings. According to their studies, there should be an offshore current, so he felt that as long as he battled the sea, he would at least be headed right.

His back and arms began to ache. Twice he caught glimpses of darker masses he felt certain were trees, and pulled the harder. He knew he was headed toward shore, at least, although now he knew that he was being sucked northwest along the coast.

Then suddenly he felt himself being heaved upward. Just as suddenly the boat dropped away from under him. The world was filled with the roar of surf. In the chaos of boat and gear he was flung down hard on the sand.

Discipline told him that he must not lie there. He must recover his gear, rifle, and get that boat out of reach of the next comber.

He did it somehow, and just before he gave way to exhaustion he was aware of his disgust for the softness of his physical condition; there would have to be a lot of hardening up if Coast Watchers were to live in this country. The cities did that to a chap.

He awakened to bright daylight. To his relief he found that he was well concealed, and so was his boat. He wriggled through some kunai grass crowning a sand bank and took his bearings.

He was within a hundred yards of a native village of a dozen or so palm-thatched huts. He decided it was best "just to bide a bit." That would give the bucks a chance to set out to work the gardens: the old folks and the children would be less inclined to be hostile.

It was midmorning when he got to his feet and boldly but slowly walked toward the cluster of huts. He had gone only a few steps when he realized that a young native girl had seen him and was standing motionless watching him. He stopped. The girl blinked once and ran toward the nearest huts.

"Now I was for it," Wright said later. "What was it to be?"

In a few moments the girl reappeared. She was leading an old man. Obviously he was a chief. Wright knew there would be many eyes peering from the huts now. This would be the test: if the old man accepted him, he would be safe, for the time being, at least. They came closer, slowly, cautiously.

Wright was smiling, and it was the same smile that had won the Yank sailor on the submarine.

The old man and the girl stopped. Wright made no move except to turn his palms outward.

The old man looked to the right, and then to the left. Then his face relaxed and he, too, smiled.

In a moment it seemed they were surrounded by the entire old and young population of the village. One woman reached out and touched his arm. She spoke excitedly to the others. The white master was no dream. He was real!

The spate of pidgin English that followed was sweet music to Wright's ears: they did not like the Japanese. Even so, the white masters had gone and it was difficult to know who was master; one had to be very careful; but they would be careful until the white master whom they liked would come back. He would come back soon, surely?

Yes, indeed, the Number-One Big Fella would come back, promised

Wright. And when he did, would they help him, just as they used to?

For answer the old chief, whose name was Nugile, issued orders and young boys darted away to assume positions as lookouts. It was a nice gesture and its significance was not lost on Wright. The old chief then led him to a hut that looked like most of its neighbors and posted inconspicuous lookouts there, too. Inside, they talked for a long time. It was plain that the old man meant to harbor him in spite of what might happen to the whole village if the Japanese learned of it.

To his relief, Wright learned that the *Chino* encampment was not far along the coast. But the chief made it plain by his expression that he cherished no high regard for his oriental neighbors. The next day Wright went to see for himself.

Nugile sent two boys as guides. At the edge of the clearing where the Orientals were living in a combination of huts and tents Wright directed the boys to wait and he advanced. He had assumed the existence of at least some displeasure, and he had been further prepared by the old chief's expression of disapproval; even so, the outright hostility he encountered was more than he expected.

His approach was viewed by two slender Chinese dressed in khaki shorts and shirts. One of them spoke over his shoulder and several other Chinese appeared. The first two stood with folded arms. Their eyes were black with animosity. Wright stopped, holding his .303 loosely.

"Anyone here speak English?" he queried pleasantly. One of the others who had approached said with an evident sneer: "At least as well as you do, I should say." Then, "And what, may I ask, brings you here? Do we assume that all the Japanese suddenly have departed?"

Wright took the measures of the man, then cast a quick glance of survey around. He could see at least fifteen Chinese in evidence now, and some of them were armed. He ignored the insult and decided to face it boldly. He started by saying that he was sure that no one he knew felt any better about the situation than they did, but that sometimes it was necessary to retreat in order to fight again. That, he added, was precisely the Allied position at the moment. He told them of the American strength building up in Australia and that the counterblow was not far off but would be the more effective with proper information in hand. Then Wright changed his tactic slightly and "tried to butter them up a bit" by expressing his delight in encountering among them one who obviously was a man of education, understanding, and . . .

But the other interrupted, finishing his sentence for him by saying:

"... like one who was educated in your own Melbourne." There was no trace of accent. He turned and spoke rapidly to the others in Cantonese. The men turned hostile glances on Wright and nodded vigorously. The leader fixed Wright with a crooked smile. "My companions do not admire you any more than I do for the craven manner in which you all abandoned us to the tender ministrations of the conquerors"—he emphasized the word—"and they suggest that we turn you over immediately to test how well you would fare with them."

Still others appeared and came forward. Wright realized he had stopped just in time, for as yet his flanks were fairly covered by jungle. Even so, it was twenty to one.

Wright said that such an idea was ruddy silly on the face of it, because by so doing they would be destroying their only chance of his being of use to them by providing the needed information to the Allies for use in the comeback that could result in the restoration of their properties.

Wright's own dark eyes locked with the black ones for a long, unblinking moment. Then the man turned and spoke again, apparently interpreting for the others. Now Wright discovered that another of the group understood English. Furthermore, he spoke a sort of pidgin. Wright made out that the man had been in Rabaul just recently. Ignoring the leader, Wright began to question him. But the leader quickly put a stop to it and admonished the man in Chinese. Wright was quick to see that his own logic must have made an impression, for the fellow accepted the leader's rebuke sullenly. He finally spoke to the leader, but he looked at Wright. Wright had become conscious of the heat all at once and realized that he was sweating profusely, but he knew he had won a reprieve.

"He has decided," the leader was saying, "that he will not give you one word of information unless you promise that in return you will take every one of us off New Britain with you when you depart."

"So, you think I shall leave?"

"You will leave as you came."

"It was my intention to stay and await our forces. I was to signal my information back by means of my wireless gear."

"Wireless? You have a radio? Where?"

"It is concealed, of course."

The Chinese from Melbourne fixed him with cold eyes. Then guessing shrewdly, he reiterated: "You will get nothing from any of us unless you arrange to take us off with you the way you came— shall we say by submarine?"

Wright appraised the group again. The conviction went through him: *I've ruddy well had it. I'll get damn-all here!* And risking a bullet in his back, he abruptly turned and stalked away. Silently his boys fell in before him.

He returned to the hut. The old chief sensed that the trip had not gone well, or possibly he knew in his wise old way that it could not be otherwise, and he did not intrude upon Wright. But that night there was a tribal ceremony and when it was over, Wright was a full-fledged member of the clan. His depression lifted from him. After all, it was quite possible that the Chinese knew little and that the leader was bluffing; it was far more valuable to the ultimate cause, then, that he do all he could to cement the loyalties of the natives.

The days that followed served to confirm at least the latter part of his conclusion. The jungle telegraph announced his coming, and one after the other of the coastal villages and even some of those inland received him solemnly and indicated their friendship.

On the day before he was scheduled to make the rendezvous with the submarine, old Nugile entered his hut. Wright had a visitor, it seemed. The old man's face was puzzled. He was a Chino. Outside was the Chinese who had been to Rabaul. He wanted to give information. But first Wright warned him that there was no "deal" because he could not make one. The man understood and Wright made notes for some minutes. It was not so much as he had hoped, as evidently the man had been prohibited from all vital areas. Still it was something, especially strength estimates of army, navy, and air personnel. Wright thanked him and promised that should it develop that he could be of direct assistance to him, he would do so. The Chinese seemed satisfied and departed.

That night the old chieftain offered material evidence of his loyalty and that of his tribe. His most treasured possessions, a perfect pig's tusk and a rare piece of bark cloth, he gave to Wright. If he lived, Wright would one day return the tusk. It was the custom, a high honor that he should be allowed its custody even momentarily.

The next night, off the dark shore, the little S-boat emerged from the sea. Wright flashed a guarded signal. They took him aboard and turned southward.

Six weeks later operators of the Australian Department of Information Monitoring Service were conning the nightly foreign broadcasts. The receivers were set on Singapore, and the announcer was intoning nasally in Cantonese:

New Britain: The British-Australian troops failed in their attempt to land on New Britain Island. Their attempt was made at night with the aid of a submarine, which has been sunk off the southern tip of the island. The troops who were captured were found to be in possession of a short-wave set with which they could communicate with their headquarters.

Bad news traveled fast—even when it was not true! The next night, while monitoring a far more distant transmitter, the Australian listeners at Canberra caught a German-controlled announcement in English:

Allied attempt to land in New Britain. A small force of British and Australian troops tried to land at a certain point in New Britain from a small submarine. The attempt was repelled. The landing party was taken prisoner, and a short-wave set was captured.

At AIB, Wright read the reports with cheerful amusement.
"Ha!" he chuckled. "The blighters failed to bag the S-boat, which did exist, and captured my radio set—which didn't. Fights a neat war on his typewriter, that chappie—sanitary, too."

Later Wright asked permission to take the tusk and the cloth to Townsville, where arrangements were made for the latter to be sent to the governor general of Australia. The tusk was suitably mounted in silver. Fate was to have it that the old chief one day would be personally thanked for his loyalty by the governor general and that the tusk would be returned to him. Wright stayed in Townsville, laboring to prepare for the organization of Coast Watchers to go into New Britain.

Although the information relative to Rabaul proper was less than had been hoped for, GHQ was relieved to know that native sympathies still appeared to be with the "white masters." Accordingly AIB received directives to insert a number of Watchers. One group would ring Rabaul. Others would be strung out along New Britain toward the still relatively secure north coast of New Guinea. The overworked and undersized S-boats were impractical for inserting such large numbers; aircraft would betray unusual activity. Feldt appealed to Commander Long. Seaworthy small ships were extremely scarce but the Australian Government came up with the *Paluma*, a sixty-foot twin Diesel, slow but steady and stable. She would sneak up the New Guinea coast from the eastern or Milne Bay end, running at night and hiding by day. Opposite the southwestern end of New Britain she would pick up her Coast Watchers, many of whom had escaped with their lives from New Britain only a few months ago and now were spoiling for counteraction. Then *Paluma* would resume her nocturnal prowling, this time depositing Watchers here and there on New Britain until she was as close to the Rabaul end as she dared go. After that she would run back.

That was the plan, but it was not to be implemented.

Before the operation could begin, the enemy suddenly lunged southward from Rabaul and struck the New Guinea north coast at two places, Gona and Buna.

Buna

HIS TACTICAL INTENTION SOON BECAME EVIDENT: the Japanese were going to use their new bases on New Guinea as springboards to stage an overland attack against Port Moresby. As previously explained, to accomplish this they actually were going to try to scale the Owen Stanley Mountains and, coming down on the reverse slopes, take Moresby from the rear.

Gone were American hopes for establishing their own north coast bases, and gone were AIB's plans for implementing the New Britain Coast Watcher plan via *Paluma*. It appeared to us generally to be a serious setback. Nevertheless, at that bleak hour it was apparent that MacArthur had his own way of assimilating bad news: it was then he issued an order for AIB to activate at the earliest practicable time its hitherto-dormant "Philippine Special Section" and to put me in charge of it with instructions to "... reestablish communications with the Philippines...."

For the moment, however, we were more than fully engaged by the problems in the Solomons and the new complications incidental to putting men into New Britain. Insertion of New Britain parties by inconspicuous small craft still seemed to be the best method. But cheeky as little *Paluma* could be, she could not defy the combined obstacle of the Japanese land, sea, and air forces to run past Buna. Feldt's "Northeast Area" Section of AIB instituted an intensive search for other small craft that under favorable weather conditions might negotiate the tricky Vitiaz Straits in short hops, then skirt the New Britain coast. There remained to be accomplished much training, equipping, and such coordination requirements as those pertaining to secure codes and radio communication.

While this was being done in Brisbane, Townsville, Port Moresby, and secret places along the north or "Rai" coast of New Guinea the enemy threat over the Owen Stanleys from Buna had in fact scaled the hump and had penetrated to the line of ridges back of Port Moresby itself before the cumulative effect of attrition stalled all forward movement and forced the Japanese to pull back. Australians, almost as exhausted as their enemies, hoisted mountain artillery pieces up sheer declivities by dismantling them. Then, reassembling them, they fired into the next jungle-covered ridges where they knew the enemy to be. The operation was repeated *ad infinitum.*

Meanwhile, in September, advance elements of American infantry had been flown from Australia to assault the Gona-Buna area frontally. They were soon followed by Australians.

Enemy planners proposed, nevertheless, to maintain their hold at Buna. They pushed their pawns out and took Lae and Salamoa, farther northwestward—thereby materially increasing the danger to the Coast Watchers assembling on the Rai coast for the embarkation attempt. Furthermore, the enemy was determined to reinforce Buna by landing additional troops and supplies. Allied air activity made it too hazardous and costly to push small surface vessels from Rabaul directly into the Buna area. Doubtless, then, he would try to land elsewhere and effect a coastal supply line under cover of the jungle. Where? Could AIB find out?

It was at this juncture that a number of excellently equipped and well-trained potential saboteurs of Lieutenant Colonel Mott's section of AIB were transferred to "Ferdinand" to accomplish primary espionage missions. These men had already been in potentially "hot" locations in Papua awaiting opportune sabotage targets. Like Feldt's men, several of them were former civil servants of one type or another or had been miners, planters, and so on. Feldt knew some personally. Thus while the Rai Coast Watchers were left free to continue preparations for New Britain, the sorely-needed newcomers went to work.

One member of this highly effective group was L. C. Noakes. Another was Lieutenant K. W. T. Bridge. Both wore the Australian Army uniform and both were wise in the ways of the mangrove swamps where the fetid New Guinea rivers debouched into the Bismarck Sea. It was these men that Feldt's assistant at Port Moresby, Navy Lieutenant J. H. Paterson, referred to as "crouched in the bush like ruddy kangaroos. . . ."

Noakes was camped on one of the few relatively dry spots at the mouth of the Mambare River. This great sluggish cloaca drained an infectious area northwest of Buna. Bridge, on the other hand, was slogging toward Salamoa, another forty miles northwest, in order to watch the enemy around that place.

For Noakes it was at first an unrewarding experience of confinement in a miserable zone of swamps where the rot of death and the fungoid thrust of new life met in the brown murk of the Mambare. Only an occasional sighting of an enemy aircraft enlivened his reports; usually the markings on the aircraft droning overhead were Allied. Then one day, as he noted a flight of several friendly planes in formation, they peeled off sharply, and he heard the sound of strafing. "They're on to something," he thought.

Noakes had been a geologist in New Guinea before the war. He was a natural woodsman and he was young and lithe. With seldom a wasted motion or a false step he worked his way up the stream through the tangled masses of mangrove roots, skirting poisonous vines wound like engorged veins around tree limbs. The planes had gone, but Noakes found their battered target. The alert pilots had seen a hint of a barge and their strafing had done considerable damage. But the real damage to the enemy lay in Noakes' discovery of the existence of a well-concealed enemy bridgehead a short distance from the barges. This was the very toehold GHQ feared, for by this time the Allies had landed assault troops by air around Buna, and while they had pinned the enemy down, the situation was precarious. Conceivably the enemy could utilize the Mambare to go inland, circle around, and cut off our men, who were daily becoming weaker through disease, casualties, and tropical exhaustion.

Noakes noted every detail in relation to landmarks that could be identified easily by Allied pilots. Then he eased out of the enemy zone. On the next Teleradio schedule he had something genuinely "hot" to send to Port Moresby. AIB in turn had something hot for air command. And air command had something hot for the embryo enemy base on the Mambare. Twin-engined Australian fighters came in and raked the concealed heart of the Japanese installation. Barges caught fire. Piled supply dumps under the palms caught fire. Tents, huts, and bomb shelters alike caught fire. Noakes slid into his former observation niche and counted casualties. There were many, but what was more interesting was the activity of the living. They were pulling out.

He was sure that they would not expose themselves in a daylight withdrawal to the coast; therefore, they must be seeking a new location to conceal what was left of their base. Noakes would be only too happy to report the forwarding address. He did. The fighters came back and smashed the new site. There seemed to be little left to move now. He so reported. But every day he made his stealthy approach, and one morning was amazed to see evidences of new arrivals and more supplies. Where was the hide? Late that day he filed another message and the next morning there was a repetition of the slaughter. This time Noakes could find no survivors. Yet he distrusted his observations. For days he searched. The newest hide was so cleverly located that although he was meticulous in his directions, the pilots had to make three sorties before the place was nullified. For a solid month this sort of thing went on—one man and his party against the determined efforts of the enemy to escape the shadow of death that followed them unerringly.

"It's the finish," Noakes announced in his communication one day. "They've skipped." He felt confident the enemy had found it too expensive. He was right. Noakes stayed in the area for a long time to make sure. He also made sure another way: he signaled Bridge to be on the watch for new activity in his area. That proved to be a shrewd surmise.

Bridge's consequent signals to Port Moresby told of a heavy new concentration at the mouth of the Waria River. This was a big, dangerous force, and would have to be dealt with at once. He described the location. Air assaults were begun immediately. The enemy already was well dug in. Nevertheless, Bridge was able to report that the constant working over was serving the purpose: it was becoming impossibly costly for the Japanese to absorb their casualties and to consolidate their base. Thus, unable to help himself here, he was unable to help Buna. And Buna was approaching a stage where it would be beyond help. Even now the emaciated defenders, who had resisted with incredible tenacity, were compelled to wear their gas masks day and night against the stench of their own piled-up dead in and around their field fortifications.

These defensive bunkers at Buna were deep under tremendous covers of felled trees and earth and were proof against anything but direct hits by bombs and the heavier fieldpieces. Their reduction was strictly the responsibility of assault troops. And let it be said that no one envied the three American and Australian divisions their job amid the stinking swamps, the miasmic half-world of Buna. Yet AIB played a side role. It was one of those left-handed, upside-down sort of jobs that would fit into no orderly training manual, no neat tables of organization, but one which, as Merle-Smith put it, "should prove meat and drink to your rogues."

Obviously the best way to drain strength from the threat that had been advancing over the Kokoda Trail toward Port Moresby was to cut off its base of nourishment at Buna. Thus added to the burdens of Allied air came a new one: the air lift of troops over the Owen Stanleys in order to take Buna from the coastal flanks. The task was so great that orders went out immediately to draft into service every available commercial airliner in Australia. Landing on cramped strips that were little more than mud, they did their job and went back for more.

But the high, thin air of the mountains and the weight of artillery, ammunition, and tanks were too much for air transport at that time. How, then, could these essentials be made available to the infantry? The answer: ships that would round the southeastern end of Papua at Milne Bay and then come along the north coast as far as possible. Logical enough. But this war was being waged in parts of the world that still knew the cannibal and the headhunters; there were only the crudest charts to show the coastal waters northwestward from Milne Bay. AIB's "Islanders" were aware that there were endless miles of reefs that would tear the bottom out of any ship, unless it ranged well northward. To do this was to invite annihilation by Japanese air and surface raiders. A distress call went out to naval hydrographers. Mapping of this character under the best of conditions was a time-consuming job. Without proper equipment, and without protection under conditions of air and sea war, it was quite possible that months and months would be required. Among other shortages painfully familiar to GHQ was that of time.

Merle-Smith called AIB. Feldt recommended that a single, inconspicuous vessel be utilized to chart lanes for ships of relatively shallow draft and low tonnage, perhaps twenty tons or so. The difference in value of many luggers of twenty tons that got through compared with any number of big ships that would never get through at all, or get there too late, was obvious, he pointed out. It made sense to Merle-Smith.

Paluma was still at Townsville. Orders went out to arm her with .50-caliber machine guns—"t' save us from havin' t' carry ruddy fishing gear when we're hungry," in the words of Lieutenant Ivan Champion, who became her commander. He knew small ships, he knew the waters, and he had those other requisites, resourcefulness and courage. It was he who had piloted the vessel that had been responsible for saving Mackenzie and many others from New Britain at the outbreak. Outsize fuel and freshwater tanks were added. *Paluma* likewise carried Teleradios and lights that could be attached to buoys. After refitting delays that brought Willoughby's hot wrath and Merle-Smith's cold wrath upon us, *Paluma* cleared Townsville with a crew that defied classification—as one American naval commander later discovered. On that occasion *Paluma* had come alongside the American officer's ship to transfer Lieutenant Commander Brooksbank, brother of that civilian Brooksbank in Melbourne, for a conference. The American officer presumed Brooksbank to be *Paluma*'s skipper. No, it was explained, the chap in the Australian airman's uniform was her skipper. The American blinked. And the fellow next to him? Oh, he was an army sergeant who was her boatswain and the device on his hat was his idea of an anchor that he had fashioned from the metal of a crashed Zero. The American officer looked at Brooksbank as if to dare him to answer his next question, which was that since her skipper was an airman and her boatswain was an army sergeant, just what was he, a proper naval officer, to *Paluma*? Managing a straight face, Brooksbank answered: "Oh, sir, I'm nothing; I'm a passenger."

By night *Paluma* moved up the coast. By day she slipped into hides made the better by a canopy of cut greens. But in part of the area to be charted her work required daylight runs. Then she became open season for all airmen, Allied and Japanese. *Paluma* herself played no favorites and when attacked she opened up with splendid impersonality on friend or foe alike with her fifties. Doubtless the preoccupation of pilots concerned with missions of a broader scope saved her life, for most of them considered her worthy of only a few "squirts" of fire—although they bothered to report her "strafed and sunk" with monotonous frequency.

Whenever *Paluma* found a reef she would place an inconspicuous buoy which could be activated with a light at night. Then she arranged for shore stations that would relay Teleradio directions to the ships that would follow the path which she was laying for them. Men from her crew would man those lonely stations. At one place Corporal L. P. V. Veale of the Australian forces, in *Paluma*'s crew, sighted an enormous reef unmarked on any existing map. All modem marine charts refer to "Veale Reef" in his honor.

The little ship put her last shore party down only fifty miles southeast of Buna. Lieutenant B. Fairfax-Ross of the Australian Army was to push on with a small party to Oro Bay and be ready with Teleradio and lights. Oro Bay one day would become a major Allied supply dump and a steppingstone to places well beyond Buna.

Paluma had survived. She had been joined by others to hasten the work, and about the time events were moving to a crisis on the other side of the Solomon Sea at Guadalcanal, AIB got Champion's signal that he was ready to smuggle through the first of the supply ships.

The ships were ready at Milne Bay, thanks to diligent scrounging by the Australian Government. *Paluma* met them. Champion boarded one of the supply vessels to act as pilot, while *Paluma* went on ahead under the command of her erstwhile engineer, Rod Marsland. The small convoy slipped out under cover of darkness and headed toward Buna.

Night after night, for more than a month, the stealthy operation was repeated as the small ships came on: converted destroyers, luggers, even captured enemy barges, laden with gunners, ammunition, and other supplies of all kinds. Of course makeshift charts dissolved in the rains; native pilots recruited to help Champion became confused; engines broke down; vessels drifted out of position and scraped coral. But they came on, more and more of them. Soon there were tanks and field guns to help the cruelly worn infantry before Buna. One Australian gunner used his twenty-five-pounder cannon as if it had been a rifle; with deadly marksmanship he sent fiery tracers straight into bunker entrances.

New Britain Toll

ALTHOUGH SHE HAD SURVIVED THESE EXPEDITIONS, it still would have been suicidal for *Paluma* to have run the Buna gantlet in order to implement the original plan of having her transport Coast Watchers to New Britain. Under Feldt, Lieutenant J. H. Paterson had coordinated things well at VIG Port Moresby and, partly through the efforts of the restless Watchers themselves still hiding out on the Rai coast, he had commandeered a small covey of launches. They were made ready to take all teams except one across the Vitiaz Strait as far as Rooke Island, where there would be a separation and further emplacement.

At Rooke Island all except Warrant Officer V. Neumann pushed on for New Britain points. Neumann pointed his launch, the best of the lot, to a tiny island called Vitu, well out in the Bismarck Sea. Captain "Blue" Harris ("Blue" because he had a fringe of red hair–but that's the Australian way of it) and his party made for a point in the Talasea area, halfway along the New Britain coast. Lieutenant Andrew Kirkwall-Smith, with an Australian sergeant named W. A. Butteris, and a former missionary named A. Obst, made for a point that would enable them to keep Cape Gloucester under view. The last launch moved toward Arawe with Lieutenant Bert Olander and Pilot Officer W. L. Tupling. The team that remained on New Guinea was composed of Captain L. Pursehouse, a former patrol officer, and Lieutenant K. H. McColl, who had been coast watching many months before when the Japanese closed in and took Rabaul, but who escaped to watch again. This pair made their way into the steamy Finschhafen area and set up on high ground where they could see the harbor.

Then the lightning struck in half-a-dozen places simultaneously.

The Japanese High Command had anticipated the Allied plan to move northwestward along the New Guinea coast as soon as Buna might be neutralized. Writing off Buna as an eventual loss (but one which would not be accepted until the last man was dead), the enemy suddenly brought in half-a-dozen convoys of fresh assault troops and landed along New Britain and New Guinea in one tremendous sweep. The Coast Watchers just moving into position found themselves surrounded by an enemy whose new savagery with the inhabitants terrified them to a point of wide-eyed agreement to help against any European.

From every hand messages as unemotional as they were dramatic came into the AIB station at Port Moresby. There was little Feldt could do except to try to warn all concerned of the latest enemy moves and to keep Brisbane advised of the rapidly closing situation. Coast Watchers had reported surface movement headed toward Finschhafen. Paterson relayed the information to Air Operations at Port Moresby and bombers and strafers went out. Transports were sunk, but enough enemy got ashore to establish a formidable base. Other convoys hit Cape Gloucester and Arawe on New Britain.

It was evident to Kirkwall-Smith and his party near Cape Gloucester that their own landing had preceded by the smallest margin a far more sinister one. How extensive was the incursion and where was its epicenter? It was vital to know, not only for reporting purposes, but for their own safety. Kirkwall-Smith decided to take three "boys" and do a reconnaissance.

The party embarked in a canoe and began a cautious skirting of the New Britain coast toward Cape Gloucester and the air landing strip there. Progress was slow. Dusk caught them and they made a careful hide. Before long they heard the sound of pinnace engines bringing ashore men and supplies. All through the long night, while they fought off the voracious mosquitoes, the sound continued. Kirkwall-Smith was not one to file an unconfirmed report. At first light he crawled through the sharp kunai grass and checked the strip. When he had counted a hundred enemy in one small area and still heard many more beyond his range of vision, he concluded that he was justified in stating positively that Cape Gloucester was occupied, "well and truly." He wriggled back to the canoe and the party slid away into mists that hung in patches along the coast. But the mist was not heavy enough to hide them from the two pinnaces that were racing toward them simultaneously from around opposite headlands. Obviously they had been detected.

Kirkwall-Smith realized that the power pinnaces would cut them off with ease. He ordered the boys to dive, and dived himself. When he came up for breath, fountains of water were spouting in even patterns all around him. The two enemy craft had opened with automatic weapons. He took a tremendous breath and went under, stroking for the shore. His clothes and his army boots weighed him down heavily, but he drove every ounce of strength into his legs and arms and kept on going until at last it came to him that he had touched bottom. He straightened up and his lungs gulped in great sweeps of air. He plunged on. When he reached the beach he fell flat. He lay there for a moment but was driven to crawl on by the crack of shots back of him. He was retching with exhaustion as he blundered into a sac-sac swamp. Around his ears came the whine of ricocheting bullets and now and then the smack of a slug into a tree. Then the firing stopped.

He was too weak to go on, and the swamp was as good a place as any when they came hunting. He waited for them, gaining strength. Slowly he began to hope that they were not coming at all. It was unbelievable. But as the hours went by, he knew that for some unexplainable reason the usually thorough enemy was not making sure this time. Or were they? Perhaps they were recruiting natives? All the day he stayed and dozed and fought off mosquitoes. Darkness came. About midnight, as near as he could judge, there suddenly came from the beach the renewed sound of pinnace engines. The enemy *had* waited there all that time for a sign of life. Slowly the engines became less distinct and finally faded. To move now would mean that he might get lost. For the rest of the night he waited in his miserable hide. At dawn, filthy and sodden, he pulled himself out and went inland until he came to a village. For a long time he considered. In his words (when he told his story later) "it was a cinch I'd cop it one way or another if I didn't get some tucker in me stomach and find out where I was. So, I went in."

The village was friendly but apprehensive. Terrible things had happened nearby. He gleaned from the natives that not long after he had left on the reconnaissance, the enemy had rushed the village where Obst and Butteris were waiting. Fortunately the Japanese had not known in which hut the men were, and both Australians had had time to leap into the bush. It had proved only a temporary lease on life. In the bush the two became separated. But both had waited until the sounds of strife in the village had ceased. Then Obst had crawled back–but the enemy was waiting for him. There were shots. Probably he had been hit. Butteris made his own way cautiously to the edge of the clearing and saw the tortures that were being applied to the helpless Obst. Butteris was unarmed, but he could stand it no longer and with great fists flaying he sent the first of the enemy spinning. It was grand and it was hopeless. ... .

Kirkwall-Smith had no illusions as to what the future could hold for him. The only chance lay in trying to rendezvous with one of the other Coast Watcher parties that still might be intact on the now thoroughly alarmed New Britain. He could not expect these natives to remain friendly. As if divining his thoughts, the New Britains signaled that he was to have food—for a journey. He took it and set out. Where? The natives mysteriously hinted of someone not too far away.

Following the directions they gave him, he stumbled, mud-caked and exhausted, into another camp. His blurred senses discerned a white man— a familiar face. It was Warrant Officer V. Neumann. But Neumann was supposed to be on Vitu Island, where he was to watch for movements out of Kavieng. Neumann explained that he had been there "until the ruddy bongs chased me out." His presence had been unpopular with them after the Japanese landings and he had been compelled to board his launch and depart. He had come to the New Britain coast in hopes of finding others of the original group.

And what of Olander and Tupling? Neumann had shaken his head with the remark that Arawe had not looked very healthy to him. Then he insisted that Kirkwall-Smith rest. After that they would make a run for Rooke Island to determine the extent of the enemy invasion wave and to try to raise Paterson for instructions.

They succeeded on both counts, miraculously managing to avoid hostile sea traffic all about them. Paterson told them by radio that he had received word from Olander that his scouts claimed Arawe to be clear and that accordingly he intended to go in. Only a short time later Paterson knew for certain that Arawe definitely was *not* clear and had signaled long and futilely to warn him. More than a year later AIB would learn of Olander's capture and that of Tupling as well, and of fates similar to those of Obst and Butteris.

It was obvious now that Blue Harris and his party would have to be recovered from the infested Talasea area. The small launch he had with him was unequal to the run which Neumann's larger and better-found vessel had made to Rooke Island. But before Feldt and Paterson could coordinate rescue actions, Harris acted on his own. He had decided that the tiny Vitu Island would be safer than New Britain, not knowing that Neumann already had left it because he believed it to be hostile. The arthritic engine of Harris's launch held across the Bismarck Sea until he beached on Vitu, then gave up the ghost. Harris soon realized that Vitu could be only a temporary haven. He radioed Port Moresby. Paterson desperately sought rescuers. He finally made contact with a Watcher named Lincoln Bell. Bell was covering the Vitiaz Straits. His area was hot because through his excellent reporting Allied airmen had just bagged two enemy destroyers. He told Paterson he could make Vitu in his launch. But now war's irony worked against him. The very bombers that had responded to his call to destroy the enemy ships came back, spotted his launch, and thinking it an enemy, strafed it on its anchorage. One of the native crew members was killed. The others were terrified. With the departure of the strafers, Bell ordered the launch made ready for sea. But the shaken natives refused. Secretly they cut the craft adrift and she piled up on coastal rocks.

On Vitu time appeared to be running out for Blue Harris.

Feldt considered turning to RAAF for a rescue attempt by Catalina. AIB had been warned not to call upon overworked RAAF except under gravest circumstances. This seemed to be the time. A message was coded to Harris to be alert to a possible aircraft alighting in a little volcanic harbor of Vitu Island. It would be a long shot. But the single plane evaded enemy air and put down at night on the waters of the little harbor. Harris paddled out quietly in a canoe but even though he recognized friends, he made no move to go aboard until he was assured that there was room not only for his party, but for his dog.

Such a man was Blue Harris, truculent outwardly, soft-hearted inwardly, uncompromisingly thoughtful of those dependent on him, man or dog. His rescue from Vitu saved him for other service on mainland New Guinea.

But that would be later. Now, on New Guinea, the only party of the November push-off to cheat the sweep of disaster was that of Pursehouse and McColl who had taken up a position above Finschhafen and coolly reported moves of the invaders. In order to supply detailed information, they often crawled on their bellies through rank Finschhafen mud and wet undergrowth to observe the airfield. Their reports were models to delight the heart of any well-trained intelligence officer. Ironically, however, little use was made of them, whereas poorly-based estimates found favor with headquarters planners. One day, though, this team would participate in a golden jackpot, for from their hide above Finschhafen they would observe the approach of an enemy armada of a score of troopships and the subsequent air attack upon it. Air observers already had spotted the convoy with troops enough to have subdued all of New Guinea. They let it come close, then struck. The reports of Pursehouse and McColl figured vitally in establishing the truth of the claim that the enemy had suffered a crushing loss through this air action. This was the Battle of the Bismarck Sea.

Gazelle Necklace

THE NEW BRITAIN LOSSES HAD CAST A PALL OVER AIB, for in a sense it was a closed brotherhood of danger, and a loss was a personal thing to every man. Yet casualties there would be—scores of them in that relatively small organization. For instance, both Bell and Pursehouse eventually would go.

Nevertheless, there was no hesitation when G2's call came for a Watcher post to be established in the very heart of that same New Britain to warn of southward planes and surface traffic reinforcing the enemy now dug in there and on the New Guinea coast. Wright would go, even though he was quite aware that in many areas the attitude of the natives had undergone a change, if not by choice. There was another young man, Peter Figgis, who had been the intelligence officer of the battalion at Rabaul. I was sent south to Melbourne to ask him. In a cold office of Victoria Barracks I put it up to him. He smiled. He'd "be along directly."

The party went out of Brisbane in an American submarine considerably larger than Wright's first craft. This was the newly arrived *Greenling*. Others were in the party. One was Lieutenant H. L. Williams of the Australian Army, quiet and serious but obviously intrigued with the novelty of the experience. As for another member he positively beamed with it all. The inside of a modem submarine slipping through wartime seas was the last place one might expect to find a full-blooded New Guinea native. But Sergeant Simogun of the New Guinea Native Constabulary had adapted himself to life within one of man's most complicated self-propelled missiles as if he had come from an old line of submariners. Simogun was like that. Big, genial, he was also a master at managing other natives—and there were three more aboard *Greenling*. They, however, were from New Britain. When the Japanese drove south from Rabaul, they had impressed many natives as workers and carriers. Some of these they had forced to accompany them to New Guinea. Wright had "rescued" this trio and now, under Simogun, they had thrown in their lot wholeheartedly with Wright; they had some old scores to settle on New Britain.

The three Australians looked mahogany-dark under the *Greenling*'s lights. At Wright's insistence they had spent what might have seemed an inordinate amount of time in serious "surfing" on the splendid sun-swept beaches south of Brisbane. Wright was thinking of his first nearly disastrous landing on New Britain. "Not that it would happen again, mind you."

The plan was for these three plus one other native to make a preliminary landing in an area called Baien. The native was indigenous to the place. This time he would be the "hostility tester." While he entered his former village, the others would stand in the shadows with ready guns should his retreat be precipitous. If he was welcomed, he would hint as to the presence of the others, and if this in turn went well...

The night of the landing was calm and black. Wright thought of that other night and felt satisfied with the whole thing. Rendezvous plans were made to enable either a total pickup or a landing of the other two natives.

Two collapsible canoes slid easily away from the *Greenling*. Her low bulk was quickly lost in the night as the narrow, cranky craft were paddled landward. This time there was no savage current to tear Wright's arms out, no confusing mist, no smashing surf, only calm, oily blackness.

Suddenly he felt himself propelled skyward. He flung out his arms but hung on to his paddle. Then he was saturated. He realized that he still was in the kayak and that the automatic reactions of training had made him do the right thing to help maintain the buoyancy of the thing. The big, silent comber had rolled under them, and beyond. It was as if the sea had sighed mightily in its slumber. There was no following wave.

Ahead of him he discerned a break in the steep foreland. This inlet was what they sought, but the village in there showed no lights. Silently they glided in and beached. They checked their guns and found them sound. The native, followed by Simogun, slid into the night toward now the barely visible huts of the darkened village. Wright and Figgis took up posts.

The minutes went by. At length Wright's vigilant eye detected the black forms of Simogun and the native as they materialized from the general mass of shadows. This was no precipitous retreat from a hostile village, and his tension gave way to relief.

Twenty minutes later they were in a closed, fetid hut in the village. The inhabitants crowded around, curious, inclined to be friendly, yet apprehensive. The white men talked to the uneasy deputy for the headman, who was away.

Apparently there were no Japanese south of the point Wright had penetrated earlier. The natives repeatedly confirmed this, then made it plain that they would like the white men to move on. Wright bargained. They could move on only if the natives provided them carriers so they could go inland. Once more careful preparation, knowledge of native ways, and a faith in themselves prevailed.

Next night the *Greenling* was awash in the darkness. Figgis paddled out and made contact. The submarine's cooperative crew loaded the rubber rafts, and Figgis' little convoy went landward while *Greenling* dissolved in the night. It was even darker than the previous night, but this time Figgis had beacon fires to guide him into the cove.

The villagers were plainly anxious for them to be gone, so vivid were their memories of the ways of the Japanese in discouraging fraternization with any European. They were no more anxious than Wright; apprehensive natives had a way of suddenly turning on the object of their uneasiness and destroying it. Consequently, the next day found them erasing every vestige of their beachhead and moving inland with a considerable safari. But with every step the carriers became less amenable to the idea of further penetration.

Figgis finally conceded that it was no use. They had made only three miles; it should have been ten. The leaders decided it might be a blessing in disguise. They would pay off and give every indication of establishing their lookout. Then they would secretly pack up and make for a good hide inland. It meant carrying all the gear themselves and it would be weeks before they could do what should have been done in a day or so. But it was the only, and safest, course.

To assure maximum observation of the coast as well as the air, a lookout was built in trees on the highest elevation in the area of Cape Orford. The soundness of their selection would be confirmed in the ultimate reports of: more than seventy hostile submarines, many with supply cargo lashed to their decks for the invasion troops on New Guinea; numerous other surface ships; in excess of a hundred air sightings.

Because of them the enemy was to know the wrath of Allied air power.

Before these logs were to be compiled, however, there would be many wet days and sodden nights, for in order to frustrate hostile observation from the air, they had made their huts under heavy foliage through which the drying sun seldom more than glinted. Nevertheless, it was the foothold GHQ sorely needed on New Britain, and on the basis of it a new and very urgent plan would eventuate for "stringing a necklace of Coast Watcher stations across the neck of the Gazelle Peninsula," just south of Rabaul. It would be a necklace that would pull tight and constrict the vital Japanese communications lines to the south. But there would be a price.

As will be seen, this prolonged period during which Wright's successful observation post operated on New Britain was one of severe strain on the whole Northeast Area organization of AIB. In New Guinea, Japanese incursions had intimidated the natives and again bitter hardship, disease, betrayal, and ambush, and sudden or agonizing death would be entered on the balance sheet in exchange for achievement, some of it truly monumental; in the Solomons, as has been described, retributive action in grim proportion to the successes that had generated it was becoming the rule.

Consequently the GHQ request for the "Gazelle Necklace" imposed formidable problems of personnel, training, and supply. I was sent to New Guinea to study the situation.

For the first time the NEA section would have no alternative but to launch a major operation without quite the usual careful preparations. (Feldt always had maintained that successful coast watching was a "venture, not an adventure.") In this case time was of the essence.

Late in September 1943 the United States submarine *Grouper* eventually landed a veritable small army of mixed natives and Australian Coast Watchers where Wright and Figgis had landed months before—that is, the submarine put them in their landing craft and started them on their way. But once more the tricky waters of New Britain were taking sides. Calling on the winds to help, they played games with the parties and tumbled them ashore, soaking. This was not serious, but carelessness owing to inexperience and haste now chalked up the first scores: it was found that no waterproof coverings for radios and field glasses had been included. This large group of some forty-three became a burden on the original Wright party, named to coordinate the deployment and operation of "Necklace," until air drops could replace the ruined gear. The delay doubtless contributed to a later disaster. But meanwhile the appearance of such a sizable group had a very heartening effect on the natives; it also increased Wright's nervousness for their combined safety. Successful drops were made and the parties began their long treks, this time assisted by many carriers. Distances of eighty to a hundred miles had to be hiked by some of the parties, with relays of carriers recruited in advance by loyal natives and especially by Sergeant Simogun. At one point, when nearly all the natives had been recruited by the enemy for road repairing, Simogun made whispered promises to them for an air raid so they could have an excuse to "go bush." The raid was dutifully carried out, the natives dutifully "went bush" – and quite as dutifully returned to repair the newly wrecked road after they had secretly portered the parties to the next relay point.

The various parties, including one led by Captain J. J. Murphy, were sliding quietly into place and setting up business. Despite everything it "looked like a go."

Then came one of those "flash" messages relayed from New Britain to Moresby and from Moresby to Brisbane:

> MURPHY PARTY AMBUSHED. MURPHY CAPTURED. CARLSON AND BARRETT KILLED. CONSIDER CODES AND POSSIBLY OTHER VITAL COAST WATCHER INFORMATION COMPROMISED.

The thing crashed upon us in the middle of the night and there followed a period of anxiety that only fast action could even partially alleviate. The use of large, uncompartmented parties had been a violation of espionage practice. Now the price was being exacted. How much was known to the enemy of the presence and locations of the others? Torture and drugs had ways of making men talk. The only course was to assume full compromise and to change all signs and other signal procedure used by the parties and to rush the altered plans north at once, otherwise other good and brave men would follow those who had gone. There were disturbing rumors about Murphy, but it soon was established that these emanated from Japanese-controlled propaganda. There was no doubt, however, that he was in their hands and that the Japanese did in fact have detailed, accurate Coast Watcher information. It would be after the war that I would follow the court-martial proceedings of a recovered if torture-scarred Captain J. J. Murphy, who confronted his one-time Japanese accuser and fully cleared his name. He had not talked. And the dead could not talk: they were Lieutenant F. A. Barrett, an officer with a fine record, and Sergeant L. T. W. Carlson, who had been Noakes's telegraphist in New Guinea. The enemy had recovered bits and pieces from the party's baggage, code books and so on, and what they lacked they had been able to surmise by torturing natives.

Bit by bit the tension subsided as stations came back on the air under the altered procedure—and lived to come on again. Even so Wright's own party at Cape Orford narrowly escaped destruction later when the enemy put pressure on native children who had seen a trace of his lookout. Forewarned, his group buried as much as they could, left some for bait, and then scrambled to a new location some miles away where they could not only watch the coastal and air lanes as was their job, but also their old camp. They saw it go up in flames, but were greatly relieved, if suspicious, when the enemy appeared to take it for granted he had accomplished his mission, and withdrew. The explanation was forthcoming from Simogun. He had wrapped a note around a stick of trade tobacco and instructed one of his natives to drop it where it could be found by friendlies. The note, apparently from one literate native to another, had said that the party had decided to call it quits and had deserted to the south where it would try to get air lifted to New Guinea. The wily old police sergeant knew his jungle telegraph. His message had reached the Japanese command.

So it was that Wright and the others stayed on, stayed to send warning after warning of movements from Rabaul. And when the time should come much later for the Americans eventually to hit Arawe, Cape Gloucester, and other coastal points in retaking most of New Britain, their assault waves would include Coast Watchers who promptly would set up their radios on the beachheads and intercept the "Necklace" warnings of enemy retaliation on the way south. It would be a beautiful repetition of the Read-Mason *et al.* story of early Guadalcanal. Day after day the enemy would dispatch heavy air formation south. Day after day their flights would be met by alerted ground fire from below and plunging death from above as Lieutenant General George Kenney's fighters would fall upon them after having had easy time to fly in from Nadzab, New Guinea, and fly high while they were doing it. The "Gazelle Necklace" was drawing tight, and those drawing it were party leaders Wright and Figgis, Captain R. I. Skinner (with him was Lieutenant John Stokie whose last name had given Mason his call letters in that other triumph), Major A. A. Roberts, and Captain C. Bates.

Imitation has been said to constitute the sincerest form of flattery. It was known that the enemy was endeavoring to emulate Coast Watcher success on New Britain with his own watchers to watch the Watchers—and also to spot Allied air flights that by this time were beginning to "paste" Rabaul with damaging blows. At Cape Orford, Figgis' operator could hear their calls. Apparently there was a "KA" party not too far away reporting to Rabaul. On the opposite coast, somewhere near where he and Wright had originally landed, was a "TA" station. For reasons unknown the TA unit faded off the air. Then one day natives hostile to the Japanese betrayed the location of the KA station and it was routed, but all except the leader escaped in the bush. The next day he died by his own hand. The natives tracked the others—as far as a yawning volcanic fissure. There they halted, yammering, for no native would willingly enter this haunt of the evil spirits. It was plain enough that here the Japanese had elected to join their own ancestral spirits. The natives recovered from the station and suicide sites records that made interesting reading to the AIB men. They perused the logs with mingled feelings, not the least of which was a fellow sympathy.

Some entries showed fair efficiency in reporting Allied vessels. In fact, the private first class who was responsible was handsomely rewarded—with one tin of cigarettes!

But then the record showed paragraphs laden with the bitter stuff so well known to our men in this solitary, fugitive sort of warfare: dwindling food supplies, increasing hostility of natives, failing health, deterioration of electrical and mechanical equipment with consequent weakening of signals.

There was one message, for instance:

WE ARE RUNNING SHORT OF DRY CELLS. WE WILL INFORM YOU ABOUT IT AT 1710, SO PLEASE RECEIVE THE CALLS.

Then as the shadows of their fate seemed to descend upon them with recognizable distinctness, there were other messages to say that they were destroying some of the equipment and burning some of the code books. Efforts apparently had been made to send relief parties from Rabaul. They never arrived. What did arrive was a message from the Eighth Army Headquarters at Rabaul:

DON'T LOSE HOPE. TRY TO GET BACK AND MAKE YOUR REPORT.

Another document had been found, a fragment of a letter from the

KA leader to his parents in Japan:

> I WILL KILL MYSELF IN AN HONORABLE MANNER, ACCORDING TO JAPANESE CUSTOM, SO DON'T WORRY ON THAT ACCOUNT.

And later:

> LIEUTENANT MORITA HAS COMMITTED SUICIDE. THE LAST TWO MEN WILL TRY TO REACH RABAUL TO MAKE REPORTS.

There was no more.

Fateful Hollandia

IN MID-1943 AND LATER, while Wright and Figgis were successfully insinuating themselves as thorns in the flanks of the Japanese on New Britain and Read and Mason were successfully evacuated from Bougainville, developments were following a darker course in the central highlands of New Guinea. Here the land had nothing in common with the steamy coastal jungles—unless it was armor-piercing mosquitoes with frightful appetites. The air was rare and could bite with cold. Trees were thinned out in the poor soil and open patches of ground offered little cover.

It was such country as this that two AIB parties had traversed to reach a position far westward near where the country was politically split into Australian and Dutch New Guinea. Beyond the border lay the prewar seat of Dutch colonial government—Hollandia. One day it would be General MacArthur's advance headquarters, but now the Japanese held it firmly, as well as all the coastal areas, and, as our people would find, even the interior. The Australian segment of the party was led by an experienced bushman, Captain H. A. J. Freyer. It was the plan to have him dig in around Vanimo, on the Papuan side, while Sergeant H. N. Stavermann of the Dutch Section infiltrated his smaller group beyond in order to establish observation on Hollandia.

Since July 1942 the Dutch segment of AIB had been experiencing one of the highest casualty rates in the whole organization. Excellent agents, some of them Dutch and some Indonesians of noble blood, had displayed utmost bravery in attempts to penetrate occupied Java and the islands of the Celebes area. Agent after agent had gone in during 1942 and 1943 and failed to show for the pickup by submarine. One, an unknown hero of a

Dutch party called "Tiger 11," had signaled a sacrifice and was heard about no more until the end of the war. In this operation the Netherlands submarine *D-12* had inserted the group into Java in November 1942 and was to make a pickup later. Surfacing the night of February 6, 1943, at the appointed place, she saw a green light on the dark shore line. This was the prearranged safety signal. Preparations were made to make the beach when Morse signals from a white light at the same shore point warned in Dutch: "Danger. Danger. Go back." The *D-12* quickly went under for obviously the party was sacrificing itself for the safety of the submarine sent to save their lives. Japanese records recovered after the war spoke of their capture immediately after the warning and their subsequent execution.

It was in such an atmosphere of desperation that Stavermann was utterly determined to succeed in the Hollandia mission.

At Nemo, fifty-five miles south of Hollandia, he established a base camp. Leaving a fine young Australian signaler, Sergeant L. G. Siffleet, at the base in company with one of the two Indonesian operatives with the party, Stavermann took the other and set off for a reconnaissance toward Hollandia.

They had been gone but a short time when Siffleet and his companion were startled by the sound of shots from the direction their chief had taken. The volume of the firing left no doubt in their minds but that Stavermann and his companion had encountered a force of much more than their own firepower. Soon natives verified Siffleet's conviction of disaster.

There appeared to be only one course of action: snatch their packs and start back along the bleak trail to the Australian party somewhere near Vanimo.

Siffleet knew the enemy would be upon them very soon. He burned those parts of the codes that flames could destroy and buried the binders of the booklets. The radio was smashed. Then the two set off on what they must surely have known was an almost impossible journey for them to make unassisted in what now had proved to be hostile country. Records taken after the war were to tell of their eventual capture. The records also would tell of an execution, that of the Indonesian who had accompanied Stavermann. He had seen his chief throw up his hands and spin around as the impact of the first hostile volley stopped him in mid-stride. The Indonesian knew that he was dead and had turned blindly to escape the gunfire. Apparently he had plunged about in a circle and, still dazed by what had happened, had come full into the arms of his pursuers.

Hollandia thus early gained for itself an evil name on the books of AIB. Numerous other Dutch parties were mounted to approach it from the south. Their efforts were splendid, but they never made it.

Meanwhile, Freyer himself was in peril, as were other Australian parties working the interior farther eastward. The interior highlands of New Guinea had been considered to be relatively safe areas. This was no longer so. The Allied victory of the Bismarck Sea was responsible for the new peril: now less secure on the coast, the enemy had gone inland. Freyer decided early that discretion was virtuous; he established native "pipelines" through which he was able to collect some information without exposing himself. Items in his gleanings made him even more dubious of his security; eventually he was shaken to learn of the fate of the Dutch party. Then one day, while in a conference with some natives he had thought friendly, he and others of his group were suddenly attacked by them. Although fairly covered by black bodies, Freyer concentrated on pulling the trigger of his gun even knowing that he could not point it. The detonation unnerved his assailants. The white men acted fast and shots and grenades enabled them to escape. It was plain, however, that existence would be a day-to-day affair. Permission was sought by radio to abandon the area. Air lifting was not practical. Before Freyer and his people would come out to where it was practical, they would have spent a total of nearly nine months under the most rugged conditions and would have tramped approximately seven hundred and fifty miles.

And yet Hollandia went unreported.

Persisting in the south with the resources left to them, the Dutch no longer had assets sufficient to mount another penetration via Vanimo. By this time, however, the Allies had implemented a tactical policy of bypassing enemy strong points on New Guinea—after having captured Buna, Lae, Salamoa, and other places needed as staging bases—and now seriously fixed eyes upon Hollandia. Under this pressure the Dutch agreed to permit an Australian-mounted party to make the effort.

It was a defiant target worthy of a daring leader.

Tailor-made for Blue Harris, said AIB's Northeast Area section.

Thus was his presence accounted for in the United States submarine *Dace* as she drummed quietly along on her Diesels through a soft New Guinea night. The moist breeze came in warm puffs from the looming mountains on her port side. The relative security of Allied holdings in Papua was far behind, for she was penetrating an area west of Hollandia.

With Harris was his second, Lieutenant R. B. Webber, together with Sergeant R. J. Cream and four others: A. B. McNicol, P. C. Jeune, J. Bunning, and G. Shortis. An Indonesian sergeant was to serve as interpreter. There were also four New Guinea police boys. The party carried concentrated rations sufficient to enable it to subsist for a short time without contacting native gardens, or the natives themselves. It had radio and would have the added assistance of a relay station secretly installed a hundred miles farther south.

Whether a change in landing plans dictated what was to follow, or merely hastened the inevitable, never will be known. In any event, the *Dace's* commander considered it unsafe to enter the area because of suspected mines and, accordingly, an alternative site some miles away was agreed upon.

The party embarked in rubber rafts and made for the dark and unrevealing coast. The *Dace* disappeared. The night was calm and quiet but there was a sub-surface swell and as they approached the shore a series of heavy waves drove them in hard.

Personnel and gear went every which way in the foaming breakers. It was a near thing for Harris. He was swept into deep water and the weight of his clothing and equipment pulled him under. But he was a strong swimmer, and he managed to make it to shore, despite the fact that he was suffering from a mild attack of malaria.

Suddenly a signal fire blazed out nearby. It was plain that they had been observed by natives—Japanese would have attacked at once. Summoning his strength, Harris spoke with them. They seemed friendly enough, but an intuition born of years as a bushman warned him that danger lay ahead. The party had lost its radios in the surf. Harris decided that they should camp nearby until first light, and then organize quickly for the journey of a hundred miles to the relay station on the Idenburg River.

At dawn they took stock: a week's rations for the whole party; four sub-machine guns; a few grenades; small arms; medicines.

The start inland was made after the natives had agreed to furnish a guide. They cleared the thick foliage of the coast and went inland, where the bald areas became frequent.

Then the guide disappeared.

It was an ominous development.

Webber was quietly told to drop behind for a reconnaissance.

Almost immediately he was back with the news that a big party of

Japanese was coming upon them rapidly.

Harris made a swift decision to split his party as the best possibility for the survival of some of them. He sent the police boys on forward where there was foliage cover.

The boys had just reached the line of cover when rapid fire burst out, seemingly from all around. Webber and Jeune took to the ground, saving their lives. One of the police boys fell backward into the clearing, dead.

Some of the eleven were destined to come out. From their reports it was certain that Harris could have reached cover and accorded himself a chance of survival. But it would have meant that the enemy would have been free to range the area for all of them. He elected to claim the enemy's attention there in the open. With him remained Bunning and Shortis. Superbly calm while automatic weapons and even mortars were brought up and directed against them, they sent an accurate, deadly fire into the Japanese and held them at bay.

Flesh and bone and raw courage could not forever balance out such hopeless odds, but they did it for four interminable hours.

By now the others had managed in one way or another to disperse enough to have a chance for survival. McNicol had tried to signal Harris to join him in a relatively secure hide but the embattled leader signaled that he should save himself.

From that point the record remained to be reconstructed from captured enemy documents.

Even the Japanese, who lost heavily in the fight, paid tribute to the warrior qualities of the man who finally faced their combined strength alone and without a single serviceable weapon. He had been wounded in three places and was almost helpless.

They had seized him and propped him against a tree. While life remained they intended to extract information about GHQ plans, about Coast Watchers, the *Dace*, and anything else that might help them deal with future parties the way they had annulled this one. It was useless: the defiance they faced was one of the spirit. The bayonets they repeatedly thrust into his dying body could not touch this—only release it.

It would be weeks and weeks later at four different places along that coast, now being assaulted by American air, ground, and naval forces, that men who were hardly more than living skeletons would tell of this battle in the interior and of their survival. The brave holding actions of Bunning, Shortis, and Harris bought the lives of five members of his party. But Hollandia had consumed the best that AIB had to offer. The miracle was that five should have survived to claw their way back slowly to even reasonable health.

In this unyielding region, then, the tactical forces had caught up with the Allied Intelligence Bureau effort—an effort that normally preceded actual assault by weeks, months, or even years. On New Guinea, GHQ's policy of "leapfrogging" had greatly speeded up the Allied timetable. While the predominantly American attack forces continued to move northwestward, tough Australian outfits took over the dangerous business of containing and reducing the Japanese Eighteenth Army still in New Guinea strong points. AIB parties that once had been riddled because of them now would come back and give great help in revealing their locations. Native battalions like those originating with Kennedy in the Solomons would perform efficiently under Coast Watcher direction.

In New Britain the Americans had pushed into all but the northern third. Others had taken points in the Admiralties. The Navy and Marines had cut communications and taken ground in the Central Pacific. Thus slowly the one-time giant of Rabaul was contained, and anemia had set in there. The cost to the Allies had been great, but not so great as if direct assaults had been undertaken.

Strategically, then, the picture was consolidating into a giant Allied pincer movement. First the combined Australian-American thrust up the length of New Guinea was to develop a northerly turn at Biak and hit Morotai south of the Philippines; then the Americans would drive straight northward with all their strength. While this was developing, the Navy and the Marines would have moved across the Central Pacific rolling up a fanatic enemy as they came.

The two widely separated jaws of the pincer would grind together in the southern Philippines.

The combined land and air command under MacArthur—once more in Manila—would then fix its eyes on the Japanese homeland, while to the south predominantly Australian strength would subdue the by-passed Celebes and Borneo. The Netherlands East Indies itself would become the responsibility of Lord Louis Mountbatten.

And what of Allied Intelligence Bureau in all of this?

As we have seen, in New Britain and New Guinea the Bureau had been vitally concerned long before our troops moved in or, for that matter, before they even had been formed up for the assault. After Hollandia, AIB would both precede and accompany the attacking waves as they rolled on to Biak, Sansapor, and on to Morotai. There, while GHQ was mainly concerned with driving on to Leyte, AIB would come somewhat more under the immediate supervision of Australian land headquarters in order to perform espionage missions in the Celebes and on Borneo.

In the north it would be revealed to the American forces as they blasted their way into the Philippines what a vital role GHQ intelligence services had been playing in the two years before their coming. They would find intelligence information pouring in from more than one hundred secret stations hidden in every part of the archipelago. They would have the benefit of advance weather warnings from beautifully complete weather reporting stations similarly hidden from the enemy. And they would have the invaluable support of thousands of knowing guerrillas under properly organized control—men whose every modern weapon, every modern piece of radio equipment, every peso of official new money backed by the United States, had been infiltrated to them by submarines cooperating with those intelligence services, of which the Bureau originally organized by Willoughby was the backbone.

How had this come about?

To trace these and other developments as stirring as any in AIB's book it is necessary to go back once more to the time when the menace of complete and final domination by Nippon lay like a black cloud over the whole Pacific world.

On that New Year's Eve of 1942, while in Brisbane the noisy crowds contrasted with the lonely vigil Read was keeping on the silent Bougainville beach, while toasts were being drunk in the blacked-out *Paluma's* cabin to a job well done and leave soon to come, while Stavermann and Freyer waited patiently in Port Moresby for an air lift toward Hollandia that never seemed to come, and while Noakes and Bridge "crouched in the bush like ruddy kangaroos" near Buna—while all this was going on, other individuals and other events that were to have solid impacts against the enemy were casting their shadows before them.

In Washington, D.C., for instance, a stocky man named Charles A. Parsons, who had recently eluded the Japanese by one of the war's best hoaxes, was enjoying a last fling with his lovely wife before he obeyed the orders in his pocket to report to AIB for assignments that were destined to make the Japanese regret that they had ever let him out of their hands.

On the Pacific side of the world, at Mott's secret training station on the Queensland coast near Cairns, toasts were being drunk by a small picked group of commandos and saboteurs to the success of the project proposed by their youthful chief to penetrate two thousand miles of hostile waters in order to blast enemy shipping in Singapore Harbor. Major Ivar Lyon stood erect in his uniform of the Gordon Highlanders and acknowledged the salute; but he was thinking of his missing wife and daughter, now thought to be dead at Japanese hands in Malaya, and he did not smile.

While Read stowed his civilian evacues aboard the submarine *Guardfish* another United States submarine far to the west was moving toward the Philippines. Sitting on the edge of a bunk in the forward torpedo room of the U.S.S. *Gudgeon* was Major Jesus Villamor. He had been named to head the first espionage party to go back into his native land.

Tinian, Bougainville. July 1945. Flight Lieutenant N.C. Sandford, Allied Intelligence Bureau, shaking hands with natives before leaving for Torokina on the motor launch ML1327, on the occasion of the rescue of Netherlands East Indies army personnel who had been prisoners of war (POWs) of the Japanese.

Tinian, Bougainville, July 1945. Japanese prisoners of war about to board the Allied Intelligence Bureau motor launch ML1327, during the rescue of Netherlands East Indies personnel from the east coast of Bouganiville.

U.S.S. *Narwhal*, the primary submarine supporting the Philippine guerrilla movement, making nine secret missions to the islands in 1943-44.

Charles "Chick" Parsons during a mission in the Philippines in his typical 'uniform.' He often wore lightweight shoes for their practicality in the jungle, and in order to be able to run quickly when needed (he had run the mile for his high school in Tennessee). He also rarely carried a weapon.

Part 4
THE PHILIPPINES

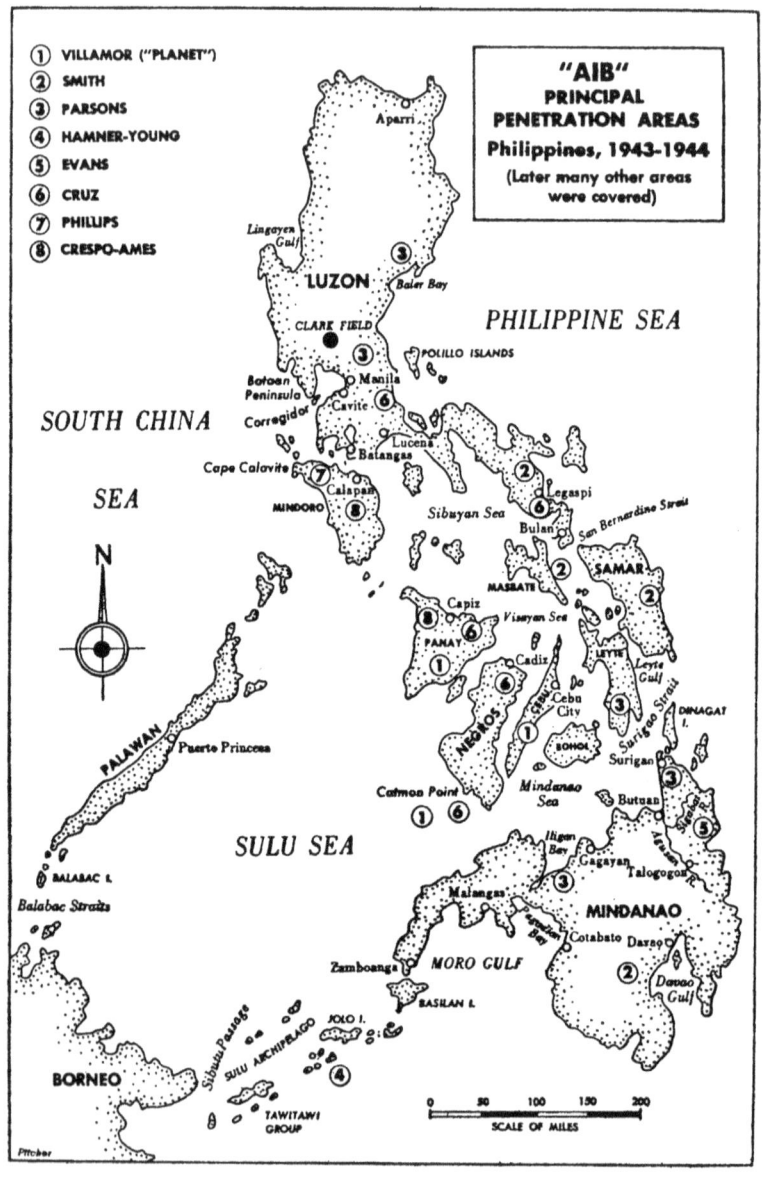

"Planet" Project

FOR VILLAMOR AND HIS ESPIONAGE ASSOCIATES aboard the *Gudgeon* that night, the previous six weeks had been a grind of concentrated training and preparations; "Planet" party would be not only the first to be dispatched by the Philippines section of the Bureau; it would be one of the best trained to go at any time from any of the sections.

As the reader will recall, the original directive from MacArthur during the dark days of Buna stated that communications would be re-established with the distant Philippines. Before "Planet" could be organized and dispatched, radio communications of a sort actually had been independently re-established. This had come about through the temerity and perseverance of some of those left behind in the Islands. Questions arose as to whether "Planet" should be canceled. G2 left little doubt in our minds: the obvious precariousness of the equipment in the Islands, the lack of secure ciphers and, above all, the contradictory nature of the information that was beginning to come out underscored the need for an observer placed there and controlled by GHQ. He must be one trained in clandestine operations and equipped with proper codes and transmitters of high reliability. The pressure was on: "Planet" went forward with tremendous energy. Even so, additional impetus unexpectedly came in the form of some incredible seafarers—tyro seamen (except one) who, in order to escape the Japanese, had voyaged all the way from the Philippines to Australia. Their feat equaled in performance and distance the historic trip of Captain Bligh in another day. There is no record of how many may have tried, but at least two parties of two men each and one of several members accomplished it. Making landfalls at different times on the northwest coast of Australia, they were picked up by RAAF and rushed to Brisbane for interrogation. They told AIB of defiant men formed into big elusive guerrilla packs in the main island of Luzon, in the middle Philippines, and most of all, in the big southernmost island of Mindanao. The enemy termed them all "bandits" and hunted them unceasingly. Some unquestionably were; and unless official recognition of their ragged, starving, almost defenseless plight were forthcoming to the others who sought to maintain a semblance of military organization, they also would be compelled to yield to the dictates of sheer survival. They knew from enemy propaganda and from a few deeply concealed short-wave receivers that occasionally could pick up San Francisco that MacArthur was carrying on from Australia. But their knowledge of developments was scant.

"Planet's" original mission of getting information on the enemy, therefore, was expanded to include bringing in token signs of hope, however minute, and of collecting data that would enable MacArthur to delegate official responsibility to the most reliable guerrilla leaders.

But who was available to personnel such a mission?

When the main Japanese invasion struck the Philippines and casualties clogged the field and base hospitals, the hospital ship *Mactan* had run the gantlet to Australia. She could not get back. Other little Filipino vessels, including the *Don Isidoro*, had reached Australia, had taken on vital loads of foods and drugs for the now-cornered Fil-American forces, and sought without convoy—for there was no such thing—to run back in. Japanese air north of Australia swept the decks into a bloody shambles and bombed her under. Some survived and were in Australia.

It was utterly suicidal to think of sending in white men: none could say how even Filipinos might be received, for none knew what might be the temper of the Philippine population by now. In that respect AIB faced the same problem Wright had faced in his original penetration of New Britain.

The only junior officer who had been selected for inclusion in the immediate MacArthur party that had come down from beleaguered Corregidor in March was Captain (later Colonel) Joseph McMicking, a scion of a wealthy Spanish-Filipino family. Born in Manila, well educated, and well traveled throughout the Islands, he had known many "contacts." He had worked with Willoughby before the outbreak on a plan for certain communication workers to go underground and form nets should the Japanese attack. (It was, in fact, one of these that had been utilized by a Major Praeger on Luzon to effect that first halting radio contact with San Francisco.) Together with Captain Allan Davidson of the Australian Army, my training officer, McMicking went on a search for a leader. He thought he knew where to find him. He was right.

Several hundred miles to the south of Brisbane, at an airfield on the outskirts of sprawling Sydney, a young Filipino in the uniform of a captain in the U.A. Air Corps had arrived at some conclusions of his own. He was convinced that he had evolved a certain aerobatic that would make him an even more formidable fighter pilot than he had been in previous combat with the Japanese over Luzon. And that was about all he lived for. He had pestered Air Operations until he got an okay to take a P-40 aloft to test his theory.

He was just ready to take off when Operations halted him: he had "an important visitor."

Important? What was more important than becoming a top-notch pilot who could do his share toward helping the Americans free his country from the heel of the invader? He felt, too, that he owed a personal debt to the United States for all she had done to give the Philippines strength and stature. Jesus Villamor was the son of a well-known jurist in the Islands. The son's patriotism was an intense, almost fanatical thing. He felt that his life was little enough to give; this was no mere maudlin sentiment—he had proved it on more than one occasion in the brief time before the overwhelming enemy air strength had wiped out the inexperienced Fil-American squadrons. In the first award ceremony of the war in Manila, General MacArthur had personally decorated him and one other pilot for their bravery. Villamor's citation was for the way he had taken off his handful of obsolete P-36 fighters from Batangas field and climbed his wholly inadequate "putt-putts" to engage a flight of not less than fifty-four enemy bombers and fighters. The

Filipinos had actually bagged one bomber before being forced out of the hopeless fight.

Now Villamor pulled himself up out of the cockpit of his P-40. He had to stretch his short legs to reach the toe slots as he let himself down.

At that moment he heard his name called. The greeting was mutually enthusiastic, for these two were old friends, as were their families. McMicking explained that he had learned of Villamor's arrival in Australia aboard one of the Royce bombers staged out of Australia in one splendid last attempt to punish the Japanese before the surrender of the Philippines.

Villamor stated that he had never expected to come out but that he had received orders to contact the Royce bombers on their way back to Australia.

McMicking looked at him for a long moment and then in an almost nonchalant manner asked Villamor if he would like to go back in.

The little Filipino answered: certainly, if he could do any good. But what was this all about? McMicking's answer was to ask him if he thought he could count on the many good friends he had had in the Islands.

Villamor was not sure. Even his own father might not be glad to see him. After all, he, McMicking—all of them—had deserted their friends in the hour of greatest need. Certainly they were *ordered* out, but... How could he know how the people up in the Islands felt about MacArthur and all the rest of them now?

McMicking said that the answer to that vital point was one Villamor himself would have to radio back. (McMicking had been so sure of his man that he had arranged with Fifth Air Force for the indefinite loan of Villamor before he had called him out of the P-40.)

Maturity forged in the furnace of war already had stamped itself upon Villamor. Coupled with that were two inherent qualities that AIB would require: an agile but comprehensive imagination, and a fund of resourcefulness absolutely essential to the undertaking that was now shaping itself in his mind. Oriental inheritance had enriched his imaginative thinking, given craft to his planning, and put a glitter in the black eyes on each side of his broad, almost conclave nasal ridge.

A few days later Captain—soon to be Major—Villamor reported as directed. His head was bursting with plans and he threw himself with amazing energy into the screening of personnel and the thousand details of planning and training. He selected men from the *Mactan* and the *Don Isidro*. The diminutive Filipino and the big, barrel-bodied Australian Captain Davidson clicked at first contact.

In his own special training setup near Brisbane's Victoria Barracks, Davidson had put his charges through a tremendous physical conditioning course. Those who survived the mornings earned the right to recuperate during afternoons of concentrated training in Morse code and cipher systems. At night there was celestial navigation and infiltration of known suburban areas without detection by an Australian population already edgy with the fear of an invasion. Then had come small-boat management, map reading, sketching, military recognition, campcraft, living off the country. First aid was followed by Judo and Judo by surf training until on a dark night they could load their landing rafts in a high sea from a simulated submarine, placing every item in the right spot without saying a word or showing a light; they then had to paddle away, make a surf landing, bury all gear and rafts, erase every track, and defy Davidson to find the cache in bright daylight.

Davidson made them chop wood and plow land in support of their cover stories that they had run for the Philippine hills when the enemy had invaded, and after farming there for a living, had come down out of curiosity to see what had happened, to see what this new "Greater East Asia Co-Prosperity Sphere" they had been hearing about actually meant. Their hands became dirt-begrimed and horny, their nails cracked and broken; their calloused feet would not now tolerate proper shoes, for few in the Islands were thought to have proper shoes.

Among the countless subjects considered during those weeks of preparation was the possible need for an emergency pickup out of the Islands. Lieutenant Colonel P. I. "Pappy" Gunn, a pioneer in Philippines aviation, had proposed flying in after Villamor should

it prove necessary. (Gunn was a man of ideas. It had been his scheme for arming Mitchells with six .50-caliber machine guns in the nose to blast off ack-ack crews on enemy transports, thus enabling the bombers to come in low for "skip bombing," that was credited with the tremendous carnage in the Battle of the Bismarck Sea.)

I had asked General Kenney about Gunn's proposal. He had no long-range Flying Fortresses left, and B-25's did not have the range for such a trip.

"General Kenney says you can't," I reported to Pappy one day when he and Villamor were in my office.

Gunn settled down on the tail of his spine, clumped both feet on my desk, and drawled:

"Does he mean that he won't let me, or that they's technicalities in the way?"

"He hasn't said he wouldn't let you—yet."

Pappy waved a hand on which the right little finger was stiff— flying mishap.

"I'll fly it there and back and take a pay load, too." His feet hit the floor. He fumbled in his shirt pocket for a stubby pencil and reached for the nearest piece of paper. It was a secret communication from G2. He was hastily provided with a substitute.

"We'll want a B-25 to start with," he said, putting down gross weight figures. "Well, now, we got to lighten her, so we take out her guns." He put down another set of figures. Villamor blinked. "Then we take out her armor plate, then her radio. See, that lightens her up, don't it?"

"But, Pappy," objected Villamor, "no guns, no armor plate, no radio. Are you taking the engines by chance?"

"We'll take 'em. Need 'em on account of we won't have no guns, no plate, and no radio. Yap. Y'see, we don't need guns because we ain't a-goin' to fight no one. We ain't a-goin' to fight on account of we got engines, good engines, and we ain't got no weight, so we'll just go too fast to bother about stopping to fight someone."

"That's clear," grinned Villamor. "I guess."

"Now, let's see, we ain't got guns, so we ain't fightin', so we don't have to carry ammunition; that's that much off again."

"The radio, Pappy? Why don't you need the radio?"

"We know where we're a-goin', the man on the ground knows where we're a-comin', and we don't want any billies back here callin' us up to go someplace else. So we don't need it. Now we ain't got bombs, so we can use the bomb bays for fuel, only we'll put in lighter rubber tanks; more weight gone. We'll fly water high only, so's we don't alert any radars, or if we do, no one can see us in time, so . . ."

Pappy paused. We held our breath. Now what would he remove – the flight deck, maybe?

"So we don't need oxygen apparatus," finished Pappy.

It went on from there, until Gunn was considering a B-25 stripped of practically all but a pair of flying engines and a board between.

"There," he summarized triumphantly, totaling his figures, adding and subtracting the weight saved from the normal, adding for extra gasoline and some pay load. Carefully computing "how much gas the ol' gal will burn" against the miles, he had enough safety factor to be seen with a good microscope. But Pappy didn't have a microscope, and didn't want one; that was the way Pappy usually flew.

"So, Jess," he drawled, "you're practically picked up."

Other problems considered during the weeks of preparation included light signals, warnings, and so on. Then alternate means of instruction should Villamor's radios fail: he would have to locate one of those hidden short-wave receivers and listen to San Francisco. Certain phrases to be repeated on certain nights. And the problem of how to smuggle in seven pages of closely typed pages comprising a high-grade cipher system for at least two selected guerrilla leaders, Fertig on Mindanao and the Filipino,

Peralta, on the island of Panay—when an itinerant Filipino would not be wearing any more covering than a thin cotton shirt, maybe shorts, and possibly a straw hat. Microfilming had solved part of the problem; altered dentistry to conceal the tiny film had done the rest in one instance, a certain clever patch on a pair of gymnasium shoes in another.

As the date for departure neared the party, even Villamor, was purposely misinformed as to the time. It was hinted that the party would be flown to Perth where Naval Task Force 77 base was located.

Instead, one day—December 27, 1942—they were issued old naval dungarees and asked to assist in loading a submarine lying in the Brisbane River. To their astonishment they saw that the covered truck was heaped with their own gear, each piece in familiar watertight tins that they themselves had helped solder, and each wrapped in innocuous burlap and hemp cord. In broad daylight they worked like any other dockside stevedores to load the *Gudgeon*. The only difference was that none of them came out of the ship as she prepared to cast off in a sudden and convenient smother of rain.

What occurred from then on Villamor described to me in vivid personal communications bearing the *Gudgeon's* own letterhead, in official reports, and in numerous personal debriefings both during and after the war. The *Gudgeon's* log entries were also productive.

It was the first time in a submarine for any of them. In the hectic days in Brisbane a thousand details of training and planning had been foreseen and provided for, yet the moment Villamor stood in that brightly lighted cylinder with its concentration of panels, dials, valves, pipes, conduits, torpedoes, and a nightmare of other things squeezing in against them from every direction, he realized that there was one item they had overlooked—susceptibility to claustrophobia. For one tense moment he had stood there, a new, strange sensation flooding through him—an urge to get out, to get back out fast into the free air and sunlight. Then he had looked around.

There was American-educated Lieutenant Delfin Cortes Yu Hico. On his fine-boned face there was an expression of eager anticipation. Beside him was an older man, heavy and solid. He was Lieutenant Emilio F. Quinto, the radio operator for "Planet." He was looking with interest at the control board with its red-and-green indicators. Those two would be all right. So would his second-in-command, Lieutenant Rodolfo C. Ignacio, who was studying the eyepiece end of the periscope with its folded handles above them. Villamor felt sure of the remaining two, Sergeant

Patricio Jorge and Sergeant Dominador Malic. Jorge had a devil-may-care expression as he stared at the jungle of machinery and then at the quizzical sailors who ran it. Malic was dependable—as he had been one day on the Pasig River in Manila when he coolly navigated the *Mactan* through amazing evasive tactics that saved her from destruction by bombs.

From the moment of their embarkation it was a new life for them in another way, for now they assumed their "cover" names and their "cover" lives. Villamor was "Ramón Hernandez" and his nickname, "Monching"; on AIB's code lists he was "W-10." Ignacio, who had distinguished himself on the *Don Isidro*, was "Carlos Noble." He also was "Brisbane" and "J-20." Quinto had been communications officer aboard the same ship; more recently he had been the chief operator for GHQ's floating wireless station off New Guinea, the *Arcturus*. She carried the bulk of the enormous radio traffic between GHQ and Washington. It had not been easy to pry Quinto away from General Spencer B. Aiken. Now he was called "Juanito del Rosario." Then Yu Hico: his new name was "Juan de Jesus" and "Dagwood," and his number, which was destined to tag many an important message being flicked down by the lightning-fast fingers of Quinto, was Z-40. Jorge became "Mr. Vicente Reyes," and Malic, "Mr. Dalmacio Canto Macilag." Each had been drilled endlessly in the details of his new life, his past, his present, and all his relatives.

The submarine rounded the tail of New Guinea at Milne Bay (where still another determined enemy attempt to make a landing for attacking Port Moresby from the east had recently been beaten back by Australians and Americans fighting up to their armpits in the swamp waters). Then she turned northwestward. It was somewhere off Buna that the ship's klaxons had sounded a strident alarm and the "Planet" men had their first exhibition of a submarine gathering herself to try for a kill.

Villamor had seen American football games and this made him think momentarily of the backfield in motion when the ball was snapped. But here, instead of four or five men reacting to an exact plan, there were scores, all within the pipelike confines of the *Gudgeon*. Even the "Planet" men had been drilled in what to do, and in one motion they rolled out of their bunks onto their feet and slammed the beds into the folded position against the hull plates. Men hurtled by, going forward. Others charged from forward and slipped into narrow slots, spinning a valve wheel or pressing a lever as they came home into their battle stations. In the control room amidships a petty officer was pulling levers that governed two banks of lights and each time he pulled, a red light went out and a green light came on. One by one the main vents to the sea were closed. The ship was rigged for diving. Now, at a command, the throbbing of the submarine's Diesels was replaced by a powerful humming; the electrics had taken over. Villamor felt the pitch of the steel deck beneath his feet. The *Gudgeon* was being driven under. He saw by his watch that exactly sixty-two seconds had elapsed since the klaxon had sounded. Already there was water over them. Now he could see how quick on the stick a pilot had to be from the time he might sight a submarine until he was in position to bomb. These guys were good!

Lieutenant Commander W. S. Post kept his crew at the top pitch of readiness; there had been constant drills since Brisbane, and now that they were in target waters, he seemed merciless.

The year 1943 was only a few nights old when the *Gudgeon* paused silently off the southern coast of the island of Negros in the lower central Philippines. She was at "neutral buoyancy"– hovering at periscope depth to enable Post to make a preliminary survey of the coast. He turned to Villamor and he was frowning. Villamor peered through the periscope and saw lights.

It could mean enemy, or it could mean only fishermen. In any event, it was obviously impossible to carry out the landing there. Some devil of ill luck seemed to be pursuing them, for already Villamor had been compelled to forego landing at the primary target site in the Pagadian Bay of Mindanao's south coast. It had been decided by GHQ that if any part of the Islands might be preserved for a future landing and base of tactical operations once the comeback had been realized, Mindanao had the best chance of survival for that purpose. Accordingly everything and anything, no matter how little, was to be done to support W. W. Fertig there. But when nosing into the bay the submarine had received radioed warnings to stay clear. Now they had missed the alternate landing point. Well, Catmon Point then, farther on. It still could not be too far across open water to Mindanao.

The next night the *Gudgeon* crept in. Post was walking the periscope around an arc in a new survey. Villamor took his turn and squinted.

This time he could discern only the darker blur of the land above a calm sea. There were no lights, although he knew from the charts previously studied that a village or barrio was not too distant.

"This is it," he said. "Glad we got everything ready. There's still time to make a landing and get covered up before dawn."

Post reported later in Brisbane that he had learned to respect Villamor. There was no shilly-shallying with him; his decisions were immediate and sound. He had been an indefatigable trainer, too. Repeatedly he had ordered "dry runs" of the landing within the restricted confines of the torpedo compartment. They had also done it in the dark.

Now the *Gudgeon* moved in closer. A final periscope survey was still favorable.

"Take her up," ordered Post quietly.

The two men were topside while water still was draining off the deck grating. With night glasses they scanned the shore.

Villamor requested that Post cover the landing with the deck gun and if he got a prearranged signal, to fire three rounds—high. By that time maybe Post would be able to discern a worth-while enemy target.

Then Ignacio came up to Villamor with bad news: One of their three rubber rafts was torn and useless. There was nothing to be gained by angry speculation as to how or when it had occurred; time was even more precious.

Post immediately offered the use of the *Gudgeon*'s wherry, the small craft every submarine carried.

Villamor considered for only a moment and then rejected the offer with "We are not trained in its use. Besides, there is not time to return it to you and we could not bury such a big object. We could muff the whole show with a slip now."

It made sense. Nor would there be time enough to make several trips between the submarine and the shore in order to take off the whole load in relays. Villamor decided to leave the guerrilla supplies aboard; everything else had to go in two rafts as had been planned and practiced.

It was a bitter decision to face up to. Villamor knew as well as anyone how desperately those hungry, ragged, sick men needed the stuff packed and sealed in the five-gallon tins: quinine, some of the new "wonder drugs," insect repellent, vitamins, surgical supplies, candy, cigarettes. And serving as stuffing around fragile phials and bottles were sheets torn from American magazines showing Japanese warships gutted by American bombs and shells at Coral Sea and Midway, and Japanese soldiers dead in the jungle—Japanese ships could be made to sink and Japanese soldiers to die, even as our own. These displays could be invaluable. Now he felt he had to leave them behind or risk the whole mission. But they had the precious radios, codes, money.

They shoved off.

Malic was on the left side of the lead raft, with Villamor behind him. Back of Villamor, Ignacio widened his stroke and got in synchrony with Malic. The raft's pace quickened. Villamor turned. In the gloom he could see only the shadow of the other raft, as there was an appreciable interval between the two.

Doubtless it was tension, but Villamor repeatedly mentioned, when he told his story much later, how impossibly long it seemed to him it was before they were able to make out the outline of the shore with its heavy crown of catmon trees for which the Point was named. The inlet he was navigating for was only about thirty yards wide and flanked on both sides by miniature headlands of tree-covered rocks. As the shore became more distinct, the submarine's form was lost to them. Now they were on their own.

They were paddling steadily and making better progress when Villamor suddenly caught sight of a gliding shadow off the right side. He turned to check the location of the second raft. There it was, well astern.

"A banca," he whispered. He was referring to a native-type boat with outriggers. Then: "Keep paddling, just as if we belonged here."

The other raft with Yu Hico, Quinto, and Jorge apparently had seen the stranger for now they lost way to observe.

Plainly the stranger had seen them, too, for he was drifting. Only one man could be discerned. But one or a dozen, the threat was real and demanded that Villamor do something to annul it. But at that moment the stranger seemed to make it unnecessary, for the shadowy craft resumed way in a southerly direction.

Villamor prodded Malic with his finger and indicated that he was to paddle again. The only concern now was what had the unknown individual been able to see and what conclusion had he arrived at when he decided to move away? Villamor tried to make out whether the stranger's pace would seem to indicate fear, or just lack of curiosity. Then his anxiety returned. The other craft was not departing at all, but had merely circled and now was silently skirting the area of Yu Hico's raft. Presently it was nearly in line astern of them.

They must entice him to come closer and see to it that he would never be able to report his observation.

Villamor whispered to Malic, the only one who spoke the dialect of this part of the Visayan area. In a loud, confident voice the reliable Malic shouted: would the stranger come to help them? They had a man who had been bitten by a shark and they needed assistance. Please.

The third shadow was motionless. Then a voice answered in Visayan: "But there are already three of you. How can only I help?"

Three? Villamor puzzled, and then the solution hit him: the stranger had seen only one raft, the trailing one; doubtless Villamor's own was sufficiently close to the shore so that its shadow was lost in the general blur of the land, and now, with the relative lineup of the three boats, even Malic's hail apparently had come from the trailing raft. The man was speaking again.

He could, he said, probably do more good by going for help. Suiting action to words, he began to ply his paddle, and the shadow of the slim banca now moved off to the right with purpose. The cumbersome rafts would be no match for the native canoe and it would only alarm the whole district to use firearms to try to stop him, even if they could sink him at that distance, in the dark. The speed of his departure made it plain enough now that he either was alarmed or was convinced of the need for help.

They landed without mishap and now began to reap the dividends of the training grind. With few false moves they buried their gear and one raft. The other raft, said Villamor, must be found if the cautious stranger should return with "help"—most likely a detachment of Japanese soldiers. Or it might be pro-Japanese Visayans. It might also be trigger-happy guerrillas. In no case was their situation good; it soon might be desperate. It was clear that he had a very important, very unpleasant decision to make. There must be a sacrifice party. The man had seen three individuals in one raft. That left three still unknown. It occurred to Villamor that he might suggest drawing lots to determine who should comprise the sacrifice party in the event the man returned with enemies who would sweep the area until they did find three men and a raft. He rejected the thought; he was the leader and he must make his choice regardless of sentiment. On this cold basis the little Filipino put down the surge of affection he felt for all of them and their loyalty and named those who must accompany him: Malic because of his linguistic facility, and Quinto because he would be needed to operate the radios.

It was done. Villamor covered the seaward approaches by establishing Yu Hico on the northern headland, while he and Jorge made their way to the southern one; the others would alert them should there be an approach from the jungle. If luck was with them and no expedition came against them, they would bury the other raft at dawn and push off landward in the general direction of a house Villamor had been able to discern in his periscope survey. But luck was not with them. Hardly had

Villamor taken his position than he saw the moving lights of many small boats.

Villamor and Jorge raced through the darkness back to the landing site. Yu Hico was summoned. Villamor spoke rapidly. He, Malic, and Quinto would hide in the jungle not far away. They would have the arms. If the approaching craft bore enemy and they were not too many, they would try to cut them down with ambush fire. Meanwhile, every man would try to cover himself because doubtless at the sound of firing the *Gudgeon* would cut loose with her deck gun, shooting high. The shelling might be enough either to frighten off the enemy or to make him go back for help. In the interval they would all try to escape into the interior—there was no thought of going back to the submarine. If it was Visayans, the sacrifice party would try to convince them that they had landed from Mindanao with orders from the guerrilla chief there to try to contact Negros guerrillas and form a consolidation.

There was no more time. The fleet of lights was now rounding the headland. They did not say anything as they separated, just shook hands.

The craft approached. With dismay Villamor noted their number. But a moment later came compensation, for against the fitful illumination as the men landed he could see that they were not in the neat jungle dress of the Japanese soldier but rather there were ragged pants, and sleeves that flapped against bare arms. Filipinos, at least. Now the nightly offshore breeze militated against his hearing what went on but it was plain to be seen that the visitors had not come to help "a man with a shark bite." A circle closed around his three men. There were arms, of sorts, and they were held menacingly. Villamor felt tears of helplessness as he saw the three "Planet" men led off via a jungle trail inland. Except for two, the remaining party boarded their bancas and went back the way they had come. Obviously the two were sentries. Could it mean that the guerrillas —for by now Villamor had concluded that such was the nature of the band—suspected the presence of more than the three they had captured?

The "Planet" men would have to await dawn and then try to escape inland. Villamor whispered this to the others and said that he would stand the watch—it would not be long until dawn now.

As the night wore on, he had time to reflect. He recalled the words of training days at AIB in Brisbane that of all the chancy games that men engaged in where the stakes were life itself, none excelled the practice of espionage for the part that sheer luck played in it. There were too many variables, too many factors that never could be fixed long enough to plan beforehand for every contingency they gave rise to. After all the months of preparation he had been compelled to abort half of his mission in abandoning the drugs because of a mischance that no one could explain, and now half his party with which he was to accomplish the remainder of the mission already was captured. Again and again he reviewed the events of the past twenty-four hours. No, he could not have foreseen these things. Now he felt the pressing sadness at the thought of his missing men. Somehow they must be saved from harm. He thanked God again that at least the captors had not been the Japanese military.

At dawn the guards stirred themselves; when it was quite light the two guerrillas looked about them, spoke together quietly, then set off. With them went some of the strain and anxiety.

Malic was awake. Villamor roused Quinto. In the submarine they had dressed themselves to be like any other Filipino itinerants: light, well-worn khaki pants and thin, soiled cotton shirts. They wore old sandals. But concealed in a very special place which weeks of training made secure (and which even now cannot be revealed), Villamor carried on his person something which he should be able to convert in order to obtain the prevailing type of Japanese military occupation currency for the Philippines. Also buried back there on the beach were some genuine peso notes, but certainly it would be unwise to flash any of that money until they were certain that it was safe to do so. They took up the faint trail, moving quietly. It was good to be in action, good to be in the Philippines again—and good to be alive.

"Halt!"

The three men froze. Out of the corner of his eye Villamor could see the muzzle of some kind of rifle pointing straight at his heart, but he could not see the man behind it. Evidently he was crouched in the undergrowth of giant ferns. There were at least two of them, maybe more. There had been no chance for him to use his .45 automatic, nor for the others to bring their heavy weapons into play— weapons they planned to bury as soon as they felt secure after landing.

"We are friends," Malic growled beside him.

"Maybe yes, maybe no," came the answer. "Put your hands behind your heads. Now where you come from? And speak truth or my gun's eye will find you."

Villamor saw the gun barrel jerk to emphasize the order. He had been thinking hard. It would be better to stick to the story they had agreed upon last night, for probably these men were from the same guerrilla force that had taken Yu Hico and the others. He formed the word "Mindanao" on his lips and Malic took his cue.

There followed a silence. The gun barrel wavered slightly. Then: "We shall see. You come with us to the boss. He is Madamba and he will kill you like *phuttt* if you are lying."

Villamor's automatic and their other weapons were taken. Now they were prodded into action. Their captors fell in slightly behind them. Presently they came to a small clearing. Several other guerrillas silently rose from the underbrush and joined them. There was rapid talk in Visayan. One of the newcomers studied Villamor intently. Then he moved aside and they started on the trail again. It grew hot. Twice more the man came alongside and stared at the "Planet" leader. Then the guerrilla stopped in front of him. Now what? wondered Villamor, and he involuntarily jerked back when the fellow's arm came up. But the other shook his head. He reached out and touched Villamor's nose, tracing a course from the middle down across his left cheek. That was also the course of a prominent scar, the memento of an automobile collision in Manila one night following a high-school party. But if the man thought he recognized him, he did not choose to reveal it then. Instead, he turned, and the trek continued.

They seemed to be moving in a wide circle, as near as Villamor could guess. To where? And who was Madamba? Pray heaven he would at least listen.

When Villamor finally confronted him, he hardly could have imagined a man less like what he had considered a guerrilla leader should resemble. They had entered a little clearing along a trail that increasingly betrayed the presence of hidden guards such as the ones who had tripped them up. Then they had entered a little house of nipa, or woven palm and bamboo, similar to some that the Melanesians constructed for the Solomons Coast Watchers. At a small table sat an oldish man with slender, delicate features. Probably a schoolmaster in other days, Villamor estimated, and he let hope rise, for he felt that here was a mind and a personality he might appeal to. The gaunt men around him, however, were obviously hostile, theirs had been a day-to-day fight for survival. An infiltration of strangers, doubtless sent by the enemy to reconnoiter them and go back with the kind of information that would bring torture and death upon them and those they sought to protect, was something they could deal with in a very final manner if Madamba but gave them the signal.

Villamor went through his story of having been sent by Fertig in Mindanao with a message of greeting and a hope for consolidation against the common foe. The guerrilla chief gave him his attention. But if he gave him credence, he did not show it. His eyes never left Villamor's except once when there were whispers in the shadows to his left.

Madamba coldly called the others around him. There was talk, and Villamor did not need Malic to tell him that it was not going well for them.

Then a Filipino approached Madamba. The old man bent an ear to his whisper. Villamor watched intently as Madamba turned, looked sharply at the man, then came forward.

For a moment he stood grim-faced before Villamor, his dark eyes fixed on him. A thousand thoughts raced across the "Planet" chief's mind. Yu Hico and the others had doubtless been captured by these same guerrillas. Had they somehow told a damagingly contradictory story? Or had he been mistaken for some enemy who resembled him?

Madamba brought his heels together and inclined his head courteously. The uncomprising lines of his features softened. It was, he said, his pleasure and his privilege to greet the famous Captain Jesus Villamor, national hero of the Philippines in her darkest hour.

It was a wild and noisy reunion a little while later as the other three "Planet" men were brought in. There were tears and shouts and backslapping, and then it spread, for the word had gone out that these were friends. For the security of the mission Villamor decided to let the illusion of their origin from Fertig's headquarters stand, except to Madamba. Intuitively he felt he could trust this mature individual whose gentlemanly characteristics were stamped upon him. Madamba's own eyes filled when he heard of the true origin of the party and that here, at last, and so unexpectedly, had come a living sign of hope.

Nevertheless, the times had taught him to be a realist. Villamor, he said, was much too well known to move about freely. It was all right here where Madamba could vouch for the loyalty of his followers, but some other time he might not be so lucky if he were recognized. Many Visayans had sold their honor to the Japanese; the danger from them was greater than from the enemy himself, for no man could say for certain when one stood among them under the guise of friendship. The "Planet" chief knew that Madamba spoke the truth but he would not at this time, at least, abandon his original idea for the penetration of Manila itself to form the heart of his espionage net. Only later, when he was talking to another loyal Visayan, one Fortuny who operated a small hostel, did he learn how dangerous it would be: on the wall back of Fortuny's head Villamor's own likeness was staring back at him; it was a cutout from the rotogravure section of a Manila newspaper dated a year previously. Fortuny noticed the expression on Villamor's face and smiled. Yes, he said, the hero's picture could be found all over those Philippine islands, pinned to the walls of nipa huts and framed against the walls of richer houses, as well. It was a "broadcast" job such as the police might have done to notify the population of a wanted man—but this time it had been a spontaneous act of admiration. Either one could produce the same all-too-final result!

Thus it was that "Planet's" headquarters was established on Negros itself. From here Villamor would send out his men to make contacts and to spot and recruit for other nets that would have only a remote connection with him so that if one of their members should prove false, or was to be taken by the enemy, he could betray only a very limited number of his associates and probably none at all of those closer to "Planet." Meanwhile, it would be necesary for him to dispatch one of the original members to the island of Panay, north of them. There a graduate of the Philippine Military Academy named Peralta had established himself as czar of the guerrillas on the island and ruled them with a heavy but quick hand. M. Peralta, Jr., was one of those whose signals had been heard in San Francisco; it had been a coincidence of the war that the one man in Washington who could readily identify him and vouch for him was on duty in War Department G2 when the messages relayed from San Francisco came in. Lieutenant Colonel J. K. Evans had been deputy G2 at Fort Santiago before the war and had helped Willoughby and McMicking in establishing the "stay-behind" nets, prior to going back to America. It had been possible for Washington to set up cipher systems of a sort, but they were "low grade." Buried on the beach by "Planet" was the high-grade cipher. Yu Hico, "Z-40," would leave soon to deliver this system to Peralta.

Villamor was told of others who had been able to send wireless signals for help. Weak and ragged and uncertain, these signals nevertheless had spanned the Pacific and the Celebes seas as well, and had been heard in San Francisco and the RAAF station in Darwin, Australia. It became clear to the "Planet" chief that they must have come from the island of Cebu, eighty miles or so across water eastward. There were, said his informants, two principal guerrilla leaders there, and many others who fancied themselves such. The guerrilla situation was as terrorizing to the population as the Japanese invaders themselves. But the most stable and reliable appeared to be a one-time mining engineer named James

Cushing. The other was a former Manila radio announcer, Harry Fenton. Apparently he was given to wild thinking and violent actions. The two had a sort of joint command in this area, the most densely populated outside of the Manila zone, and again, except for Manila, the most important to the Japanese militarily. Villamor felt that it was of utmost importance to determine the truth of the Cebu situation. But he hesitated to send one of his own men. It was quite possible that the hard-pressed leaders, good or bad, would accord his messenger, a stranger to Cebu, the same welcome Madamba had felt obliged to give the "Planet" men.

The invaluable Madamba and another man named Castaneda spoke of one Dr. H. R. Bell, a member of the staff of an American school in the Islands. Bell had been teaching physics at Silliman University when the invasion occurred. He had buried the essentials of a radio set in the hills. Later he had reconstructed it; this was the "collection of razor blades and old bottles," as it was jokingly referred to later, that had spanned oceans. Bell was hiding out somewhere in the nearby hills, as were other members of the University staff, men, women, and children alike. (If they could only be evacuated somehow before the enemy finally did catch up with them.)

Villamor acted. Bell was located and they had a long, earnest conversation. The Silliman teacher obviously was much affected by this link with MacArthur and set off on his parlous journey for Cebu. He was to report later in confirmation of the worst that had been said of Fenton, as well as the good of Cushing. Fenton, he said, appeared to have become mentally unhinged. He laughingly related how he had rather too-narrowly escaped death at Fenton's hands when the man threatened to shoot him dead on the spot—and Bell thought he certainly meant to do it. Villamor sent an agent recruit to the Cebu area to keep Fenton under observation. He was a quiet Visayan named Abila.

Meanwhile, it was believed advisable to establish "Planet's" secret command post farther north in order to be closer to Luzon. The location of the radio station was left to Malic. So it developed that an itinerant seller of bananas trudged the area of higher ground around Cartagena, near Sipalay in northwestern Negros Occidental. One day he sat down to rest near an old nipa hut whose original owner apparently had been partial to the sea to the south and had built on a high point. Fortunately there was no trouble in gaining possession of it. A transmitter especially built for "Planet" by the United States Signal Corps unit in Melbourne was brought in during the night and carefully concealed. The aerial was a masterpiece of inconspicuousness. Tests showed that the unit was efficient. From Villamor came the word that the daytime bananaman should make the attempt to call up KAZ, Darwin.

On the afternoon of January 26, 1943, the peddler of bananas toiled wearily up the trail of broken country toward the nipa hut. Frequendy he stopped to rest his short, stocky body. And Filipino-like, he was not averse to a siesta in the shade of a catmon tree. It was a pleasant afternoon and there were few sounds save the whirring of insects, the occasional squawk of a cockatoo, or the sudden quipping of monkeys swinging in graceful arcs in the taller rain trees. Quinto was actually killing time until night, when he knew the range of his transmitter would be at least three times its daylight span. He would need it all. Far over the calm sea to the south he thought he could discern the blue smudge of Mindanao. Far beyond that would be the Celebes group. Then Timor and finally northern Australia itself. The transmitter was good, he knew that; it would develop fifty watts of output, considerably more than an Australian 3BZ, the Teleradio. But the distance was great.

The sun went down, the brief dusk settled into night. Now he was in the hut. With only the faintest dial light to guide him, he began to work his fingers over the key, calling KAZ and signing "4E7." He sent only short "squirts" because he knew that everywhere there were enemy monitors, trained men with good gear. And his was a new signal on the air. He called again. And again. Each time he would alternate with listening, rocking the receiver dials gently.

And then he stiffened, and a slow smile showed on his face.

He turned to the key again and his sending was very fast.

Contact!

Doubtless Villamor had considered how anxious we in Brisbane might be about their welfare, but it was characteristic of this man to proceed with utmost caution, one solid step at a time. And now he felt sure. But for all of us in AIB the month of January 1943 had become endless. Our troubles in Bougainville and in New Britain had the dubious virtue of compelling so much of our attention on them that we could not worry exclusively about what had happened to *W-10*. *Gudgeon's* radioed word that she had landed the party and had gone on her business had only heightened the mystery, for there had been ample time, we felt, for Villamor to have reported. We began to dread calls of inquiry from Willoughby, or from General Richard K. Sutherland, the chief of staff.

On the afternoon of January 27 I received a summons from G2. I expected the worst. But one look at Willoughby reassured me. The message was brief. Conditions were bad, it said among other short sentences. But it was *contact*. GHQ now had its own eyes and its own ears in the Philippines once more. Together we trooped into the chief of staff's office. Sutherland was a man of few words. He read the message with evident satisfaction, however, then said: "The first to go in, but not the last. We're on our way!" Later the Commander in Chief sent his congratulations—and a score of special requirements for information.

The good news was a tremendous lift for us all. It had come at a particularly useful time—the end of January 1943—for by then we were deeply committed to the preparation of another party designed primarily for the support of Fertig in Mindanao, and up to then, as with "Planet," we had been working in the dark as to the efficacy of some of our methods. Now we had reassurance. While it was becoming apparent that there likely would be a shift of emphasis to the supply of guerrillas and the utilization of intelligence information gathered by them, it was still essential that there be "neutral" observers and that they be able to report through their own secret channels. We would know one day—but not until the end of the war when on-the-ground studies could be made—that the pioneer work done by Villamor was a standout. His careful, deliberate planning and his rigid adherence to the requirements of security for such operations were sound. In the months following the landing, he literally "dug in," for he believed that he would be in the Islands until the end of the war. He recruited with utmost care. He surrounded himself with utterly trustworthy agents. He even got married.

He was not spectacular, not theatrical, his pen was not brilliant, but sincerity radiated from every fiber of him. Mistakes? Certainly: the inevitable by-product of men of action.

Now came the time for the dispatch of Yu Hico to Panay.

Villamor had wanted to go with him, at least part of the way. But now he was aware that one as well known as himself should not be seen in the company of the man who would have to traverse the unknown country to the north and west. Panay, like Cebu, was dangerous territory. Between Negros and Panay were the Guimaras Straits. The Straits were a favorite haunt of patrolling vessels whose flag was the Rising Sun and whose guns were quite prepared to emphasize the fact.

So it came about that some days later Villamor was waiting behind a clump of foliage at the beach line. His eyes were upon the slender Visayan with sharp button eyes who approached barefoot along the sandy strip. A pair of well-worn gymnasium shoes was tied by their strings around his neck. Time seemed to mean little to him, and twice he stopped to throw driftwood into the miniature breakers that curled onto the beach. Then he settled down on the sand, Filipino fashion, squatting, while he deftly split a coconut with his machete. Slowly he let the white milk drain down his throat.

"I can hear, sir," he said quietly, apparently to no one. "It is good we decided on this way; there were people—it was as if they suspected something for they followed me. But now I think they are gone."

Villamor looked down the beach. No one was in sight.

Villamor asked if he had "it." Yu Hico replied that the miniaturized cipher system was safe "where they had arranged"—sewed in the ankle patch of the gym shoes.

The "Planet" leader warned him that he had a very hard trip ahead of him but that everything was depending on him because Peralta was becoming more restive every day; it was quite possible that, feeling desperate at the seeming indifference of GHQ to his plight, he would determine upon some independent action which could easily lose all of Panay to the strong enemy. It was agreed that they might not see each other for six weeks or so. Then Yu Hico was to report back to that very spot on the beach. Someone would be waiting who would know the other half of an identification phrase and he would bring Yu Hico to Villamor's hide.

In his turn Yu Hico pleaded with him to stay hidden, as he had "seen many pictures in the houses." Villamor recognized the peril of this but said that GHQ had directed him to make a trip to Mindanao. Furthermore, it was necessary to develop a strong Cebu grapevine and at the same time push other tentacles toward Manila. He turned the talk back to Yu Hico and adjured him to remember that if he ever contacted another approved agent, the first order of procedure would be agree on what it was they were purporting to talk about, inasmuch as it was enemy policy immediately to separate persons talking together, then quiz them on the topic of their conversation. If there was failure of the two stories to agree . . . "Well, you would not come out of it, my friend," finished Villamor. *"Mabuhay!"*

The "itinerant" Filipino finished his coconut, tossed the remains into the thicket, and sauntered on—on into long miles that would be rendered as dangerous for him by the guerrillas as by the Japanese military because he was a member of neither. (But one day the message would come down in Yu Hico's own cipher: "Mission accomplished.")

Villamor made his way back to his hide-out to commence plans for the trip to Mindanao as directed by G2. The message had been among the first of the exchanges that now constituted the regular "skeds" with 4E7. G2 wanted more information about conditions on Mindanao.

Before he could start a runner reached him from the east. There had been a find in Cebu that looked highly important—a code book perhaps—retrieved from a beached Japanese naval vessel. There were no other particulars, but Villamor knew this one could not wait; Brisbane must have it soon. What to do? He must go south. Why not, then, have one of his recruits now in Cebu—a reliable agent named Alvaraz—bring the precious find to Mindanao and give it to him there? He acted upon this plan, giving Quinto the data and asking for a submarine to pick up the find from an agent in Mindanao's Pagadian Bay area.

Then he pushed off in a baroto, a larger native-type boat. Two of the remaining "Planet" men were with him.

It was a clear day and a good wind helped them along. But it seemed a snail's pace to Villamor, who with the others constantly scanned the horizons for the sight of a hostile patrol.

It was as if their alertness and expectancy actually had proved magnetic, for sure enough a dot in the west rapidly grew in size as it bore down on them.

Fascinated, they watched. They were too distant from land to try to swim for it and the oncoming craft obviously had the speed to overtake them.

Villamor made a quick decision to turn back before their original course would be plain to the other craft. The Negros shore was considerably the closer and if the examination did not prove to be too searching, they might possibly deceive the stranger into thinking that they were merely fishermen or traders.

They could see the Japanese flag now. The sight of it did something to Villamor. He said later: "It just plain made me sick—I mean sick. So I lay down. I should be ashamed to admit it but this was mistaken by the others as a sign of supreme contempt for danger. And suddenly all of the tension seemed to drop away from them and they in turn assumed attitudes as if they also were carefree and indifferent. I honestly think this miserable business of mine saved our lives, because the enemy patrol came in close, took a good look at all of us seeming to be enjoying ourselves just languishing on a hot afternoon, and suddenly they sheered off to the south without asking us a single question. Me a hero?"

But there was no doubt now that they would have to continue back toward Negros, for the enemy vessel appeared to be in no hurry to quit the area to the south; until dusk they could see her cruising this way and that.

So it happened that Villamor was not on Mindanao to receive his Cebu courier. Alvaraz arrived and delivered his precious package to Fertig's men. In due time it came to Brisbane.

It was in fact a Japanese naval cipher system, the first we had taken and one which, together with another destined to be taken in the same area later, had vital bearing on the showdown naval engagements of Leyte Gulf in late 1944, the greatest naval battles of the war.

On Negros, Villamor received a visit from the mystery man of Cebu, Cushing himself. Among other matters, he wanted to talk about Fenton.

Fenton had become a serious problem. He had smuggled out of his place of former employment in Manila enough radio parts to enable him to erect a short-wave broadcasting station sufficiently powerful to reach not only his enraged Japanese listeners but San Francisco as well. His statements were bitter and potent and certain to bring down upon him and those around him strong retaliatory action. But more to the point was that in his wild, uncontrolled manner he was shouting to the enemy much information that could prove hurtful to the guerrillas and to innocent civilians alike. On instructions from Washington the San Francisco station repeatedly broadcast instructions for him to refrain forthwith. If he heard them he gave no sign, indeed his anti-Japanese broadcasts increased in frequency and vituperativeness. Cushing said that he had felt obliged to put Fenton under arrest during his own absence on this trip to Villamor, especially because Fenton's use of firearms was unpredictable and sometimes lethal. They discussed possible solutions to the problem, unaware that at that very time the matter had been taken out of their hands permanently.

One of Villamor's watchers had been spending his days squatting before a nipa hut on the outskirts of Barrio Maslog on Cebu in which there moved unkempt guerrillas armed with knives, clubs, and some *paltics* (homemade hand guns, the basis of which usually was a length of gas pipe serving as a barrel). He had often conversed with villagers who also kept uneasy watch on the house, and he knew that the civilians lived in terror of the man Fenton. "He is mad," they would say. "He will bring the angry *Hapon* [Japanese] down upon us."

But it was obvious that something new was developing and now there was no talk, only unblinking surveillance of the house. The Filipino lieutenant whom Cushing had instructed to place Fenton in custody had taken matters into his own hands. A drumhead court-martial was in progress. Fenton was the defendant on a charge of willful murder of a Catholic padre who had irritated him. The proceedings were terse and soon completed. Fenton was adjudged guilty and ordered to be shot. The execution was carried out.

Villamor's man had heard and seen enough to go back to station 4E7 on Negros with his report. In the excitement he slipped out of the barrio and headed west. His decision probably saved his life: that night in the mountains he could see the pulsating glow in the sky. The jungle telegraph carried the story of the sudden descent of Japanese raiders determined on revenge for the humiliations they had suffered at the hands of the man whose death at the hands of his own men a short time before had cheated them out of personal satisfaction. But they were not to be frustrated completely. The barrio went up in flames amid the cries of the dying.

In his conferences with Villamor, Cushing had displayed those characteristics which prompted the former to send messages urging the recognition of Cushing as the Cebu leader and asking for money and extra radios to be sent by the next submarine in order to supply him with a direct channel to GHQ. Meanwhile Villamor proposed to provide Cushing with a small Australian Coast Watcher ATR4A transceiver to enable him to tie in to 4E7 or perhaps Fertig's local station. AIB replied that if Villamor could spare an ATR4A, he should do so as the submarine *Thresher* would call presently and provide him with more radios and supplies. So it was arranged that Cushing's little party took back one of those superb Australian transceivers which was comparable in size to the average household breadbox, but which under good conditions regularly gave service of more than a hundred miles on voice and two to three times that on telegraph. (Prototypes were sent to America for "exact reproduction" to help us supply the demand; once more, but not for the first or the last time in the war, we were to be balked by that American characteristic which insists that "bigger" and "better" are synonymous terms—the "improved" product that came back was a splendid piece of equipment, but so bulky that it required jeeps with trailers and fine, surfaced roads as incidental accessories.)

By now Villamor had decided that the original site near Catmon Point was superior and he moved back there. Ignacio, Malic, and he occupied a nipa hut in an isolated area, access to which was made difficult by both natural obstructions and phantomlike guerrilla guards with quick eyes and flashing bolo blades—bolos were silent and they saved precious homemade ammunition.

Before he left the north, however, Villamor had made solid arrangements with one Major Ricardo L. Benedicto for the establishment of a "forwarding station" to which agents going north might report and those going south could make contact and file preliminary radio reports. To enable Benedicto to maintain communication and eventually control a local radio net, Villamor named Ignacio to smuggle a transmitter to him. It was done. Ignacio turned back, meanwhile congratulating himself on the success of his mission. It was premature. On a narrow road he was approached by a man who obviously recognized him. He seemed almost tearfully joyful. The man was his own cousin.

"But, *amigo*, there is some astonishing mistake," Ignacio heard himself saying calmly as he backed off. "I have only now set eyes on you for the first time. Excuse, please."

His cousin gaped. "Not Rudolfo!" He shook his head, asked pardon, and turned away, mumbling at the extraordinary resemblance.

While Ignacio was making his way back, two more AIB submarine parties landed farther north on Panay with radios and drugs. They were *Peleven* under Lieutenant T. Crespo and *Peleven Relief* under Lieutenant Ireno Ames. Watcher stations would sprout.

But now Villamor prodded deep. Into the very heart of Manila went a powerful transmitter. This was the sweet essence of daring. To do it he sought his old friend, Castanega, who was a trained radioman.

"Memorize the circuit, *amigo*," said Villamor. "Then heat your soldering iron."

It was a curious sight, but few saw it—baskets of *camotes*, potatoes, of coconuts, and corn. In the hearts of the baskets and sometimes in the hearts of the vegetables themselves went individual parts of the radio set, scores of them. Similarly buried in the brain of Castanega was the method of making a cipher system.

"*Mabuhay!*" Villamor waved to the innocent-looking vegetable boats putting off for the north. God willing, they would negotiate the heavily patrolled waters and eventually land on southern Luzon. Then into one of the many broken-backed Filipino trucks, or perhaps several carts, would be loaded the baskets, and they would groan off toward Manila along the provincial roads with their powder of white dust, their shadowy huts and stilted houses—and their enemy road blocks. He could hear his agent protesting: But *these* vegetables, honorable sairs, were for the honorable Japanese officers' tables only, sairs; surely they must not be delayed, no? No, they would not be delayed. They were not delayed. Doubtless some of the provender did go to the tables of Japanese officers—less the technical trimmings. Those trimmings found their way one by one to a certain house within the ancient Walled City of Manila. There an American named Frank Jones, whose father was an American and whose mother was a Filipina, was indeed happy to receive them. He assembled them. Eventually he operated the reconstituted unit with great effectiveness. He was the brother of Helen Jones, who was herself doing a tremendous task north of Manila in maintaining a "forwarding station" to pass American agents within the very shadow of the enemy-operated, stone-girt jail at Malalos. Jones' Intramuros station would be the original wireless contact for many of the Luzon guerrillas.

The "Man-Who-Walks-Like-a-Ghost"

IN MID-JUNE OF 1943, WHILE VILLAMOR was engaged in developing his organization for penetrating Manila, he was both mystified and made apprehensive by a message from Brisbane advising him to expect a certain Manila-bound "Major Suylan" on the next upbound submarine. In the complexities of W-lO's personal cipher it emerged that "Suylan" actually was Dr. Emidgio C. Cruz, the personal private physician to the Philippines' president-in-exile, Dr. Manuel Quezon.

Restive in Washington under the burden of frustration concerning any effort that he might personally make from Washington for the relief of his beloved land and his people, President Quezon meditated continuously on the fate of those in Manila who had been close to him. A realist, he knew well that the forceful retaking of the Islands would not be achieved for a long time. Yet, could he not be of some service? Only those whom he utterly trusted, and whom he prayed might still live, could tell him. The American intelligence effort, he knew, was as fully developed as it could be for those early days. He was aware of Villamor's assignment to "Planet" and personally approved of him and the over-all plan. He did not desire to interfere with GHQ's operations or to suggest complicating the missions. Therefore, he proposed sending in one of his own trusted staff, if GHQ would only assist as much as possible in forwarding him and bringing him out again after he had made contact in Manila itself with one man whom he felt capable of giving him the most comprehensive, most objective report of the situation there—General Manuel Roxas. There was reason to believe that this beloved leader and his family had been put on parole by the Japanese, although he had steadily refused to enter into any agreements with them or to give his word.

So it developed in July of 1943 that a slight, quick-eyed, quickbrained Filipino civilian arrived by air one night in Brisbane and went directly to the Commander in Chief. So fantastic did it seem to General MacArthur that a man totally untrained in espionage work might effect such a daring penetration that he told Dr. Cruz in all frankness that he considered his chances of getting in not greater than 10 per cent, and of getting out—"none at all."

The submarine *Thresher* was going in the next day with additional radios and other equipment and supplies for "Planet." AIB was directed to see that Dr. Cruz, actually a lieutenant colonel in the Philippine Army and a prewar friend of Villamor, was flown to Perth for embarkation. He was to report to Villamor for further instructions.

At his end, Villamor was advised by radio to assist the forwarding of Cruz in every way consistent with the security of his own mission. That last qualification was important, as there was no intention of risking compromise of "Planet" or any part of it on an assignment which was fairly bristling with risk; in no sense was it considered to have involved the "calculated risk" of planned espionage, but, if anything, calculated demise! Villamor so estimated it when Cruz reported to him, "complete with the pocket litter of a comfortably established, peacetime Filipino civilian, including a modern fountain pen of a popular American brand that in all probability had never been distributed in the Islands prior to the outbreak of the war, and worse, at a time when the great, ragged majority of the Visayan population long since had been relieved of such personal possessions by the enemy." Villamor was dismayed that AIB should have placed him and "Planet" in such a compromising position and he took immediate steps to isolate Cruz from "Planet" operations.

To aggravate this aspect of the situation, the heart of Cruz, the physician, immediately was touched by the plight of the wretched people he saw on every hand; forthwith he announced his determination to do what he could to relieve their illnesses. He thus proved himself a laudable disciple of Aesculapius but a dismal exponent indeed of clandestine *modus operandi*. Word of the healer spread rapidly among those who so greatly required the healing arts—perhaps to others as well—through the jungle and across the blue straits.

Shortly there appeared from the south a Filipino who was not a Visayan. He was Abduraman Ali, a Moro tribesman. Villamor made another appeal to Cruz: Cruz acted on the warning and was discreet. Silently as he had come, the Moro remained, asking no questions but seeing everything. Villamor "put a tail on him" day and night—and had to admit in the end that he could *prove* nothing, yet in view of what occurred
. . .

Three days after the *Thresher's* departure the enemy landed on the beach. "Planet" was dispersed and 4E7 went off the air. The raiders knew precisely where to find the precious supplies. No personnel were taken. The enemy departed.

Thoroughly alarmed now, Villamor rightly or wrongly considered Cruz the prime factor and thought him fully compromised. In accordance with the dictates of good security for "Planet," he refused to allow his agents to associate with Cruz in forwarding operations. Cruz departed and managed to get to Panay largely by his own efforts, only to be blocked by Villamor's man, Yu Hico, from making contact with Peralta. Meanwhile, Abduraman Ali eluded his "tail" and was seen no more on Negros.

It was the beginning of a taut, unpleasant situation that may well have been responsible for a measure of the criticism eventually leveled against Villamor. Personally, I felt that his actions were justified—there was much more at stake than this heavily-loaded mission by Dr. Cruz—even if it should work out.

Cruz returned to Negros, and after considerable hardship made contact with the fugitive governor of Negros at a hidden mountain retreat well up the side of a volcano cone. Here Governor Montelibano arranged for appropriate clothing and a guide to take him through Japanese-held towns in order to reach Cadiz, on the northeastern coast, where he was to procure a sailing vinta in which he could negotiate the inland seas toward southern Luzon. Significantly, before these plans could be implemented, the Japanese attacked and drove the governor's party still deeper into the hills.

Cruz laid in a peddler's stock of dried fish and chickens. From island to island the little craft went, Cruz buying and selling the while. The vinta inched northward.

One hot afternoon a Japanese patrol boat came up and drew alongside. Her officer and interpreters boarded. The searchers dug among the dried fish but failed to uncover the waterproof packet of letters Cruz was carrying from the Presidente to General Roxas and others. And it still might have gone badly had the search not eased when the officer became interested in interrogating him. Information was wanted, he was told, about a certain Major Suylan who had brought arms to Negros aboard a submarine.

Cruz was made to go over his story of being a peddler again and again. Then the interpreter said to the officer: "He is no Tagalog, by his speech he is a Visayan." Cruz blessed his facility with dialects. From this he concluded that they really sought a Tagalog, one of the peoples from the Manila area of Luzon. He was a Tagalog, but he never would speak Tagalog on this trip. The patrol officers released him, and he reached the Bicol area of southeastern Luzon. Here he paid off his crew and slipped away. Shortly thereafter two hundred pesos procured him certain documents, including a pass in Japanese and a resident certificate for one "Emilio Corde," trader. He set off for the town of Matnog, where, if he was lucky, he should be able to find a boat of some sort bound for the area south of Manila itself.

Cruz was lucky. Amiability in the shape of one Chinese merchant named Tiong Hing awaited him. Hing consented to a proposal that they represent themselves as trading partners en route to Manila to sell their produce. The next day they loaded the sailboat with lumber, firewood, and cassava flour. They headed straight for the Japanese patrol-boat base at Bulan to sell their produce. Tiong Hing was nervous but Cruz had become hard forged in the fires of danger. They went in.

At the wharf Japanese constabulary accorded them an almost cursory inspection. Cruz was still blessing their luck when two soldiers with fixed bayonets stepped up. They would, they were informed, follow the soldiers into the town.

Stunned by this new turn, Cruz was slow to realize why Japanese and Filipino flags were flying everywhere and the people seemed to be in holiday dress. Children carried miniature Japanese flags.

Then it came to him. The date was October 14, 1943, the inauguration day of the Japanese Puppet Philippine Independence. He thought bitterly: What an appropriate time for an execution! They approached the public plaza, packed with people and Japanese soldiers.

To the amazement of Cruz and Hing, the two armed soldiers now abruptly turned away and left them standing—enforced spectators to a parade! Like the other captive spectators, they found it expedient to take their cues: they cheered when they were directed to cheer and bowed when it was indicated that they should bow. There was more speechmaking and finally three heavy "Banzai" cheers led by a puppet on the platform.

Now they were free to roam. They did—into an outright compromise. Apparendy the man had been watching them during the formal part of the ceremony. Now he mounted his bicycle and pursued them. Wheeling around two Japanese policemen, he suddenly leaped of! shouting: "Cruz ... Cruz ... I am delighted to see you, my old friend. And they told me that you were dead. Your mustache almost had me fooled, but..

Cruz engaged his old classmate, Dr. Castro, in a bear hug of "affection" that drove the breath from his body and left him speechless. He maneuvered the man around a corner. Then he explained and left Castro in open-mouthed amazement.

Soon they were aboard again. For two days all went quietly. Then another patrol approached, altered course, and made for them. Cruz had a premonition that this time it was going to be close. He took the bundle of letters he had protected all the way from Washington. Besides those from the Presidente to General Roxas and others there was one from Vice-President Osmena to Mrs. Osmena, still in the enemy-held Islands. He weighted the packet with a rock and slipped it over the side.

The patrol drew up. Through an interpreter they were told to crawl upon the outriggers of their boat and there kneel while the search was conducted.

Cruz forgot his aching knees in his thankfulness that he had jettisoned the letters. Then came the interrogation. It was thorough and again there was the emphasis on "Major Suylan."

"The honorable officer says that they know he is not far off," explained the interpreter. "Always now they are close upon him, but for the time being he is fleet. He is known as 'the-man-who-walks-like-a-ghost.'"

Cruz stuck to his story with success. He even got a note of introduction to the Japanese officers living in the New Banahaw Hotel of Lucena who would buy most of his provender. Lucena was his objective.

It was on October 22 that his new-found friends at Lucena bade him farewell at the railroad station in order that he might go to Manila to further his amiable dealings with the Japanese there. The commandant had given him a letter to facilitate this. He said good-by to his estimable Chinese partner.

Aboard train the guards gave his scanty baggage only a cursory examination. He arrived at Blumentrit Station and lost himself among the crowd. It hurt him to see the patched clothing, the shoeless feet, the tired, spiridess faces of those about him. But again he fought down every urge even to speak to anyone and unostentatiously he went along the war-battered streets until he came to a certain house. His heart had seemed suddenly too big for his body and he felt breathless as he entered the old, familiar garden of his sister-in-law, for now he would know what had been the fate of his own wife.

The woman's hand flew to her mouth, but she controlled herself. Yes, yes, Mrs. Cruz was alive and reasonably well, although there had been frightful happenings. And he must be so very careful, for the Japanese did not believe that he had died on Bataan and they were still seeking.

The reaction of Mrs. Cruz was different. The shock made her knees collapse and she dropped before the image of the Holy Virgin Mother. For a long time she prayed, slowly gathering courage to look around again to see if the man she had almost given up for dead actually was there.

This could be no prolonged visit. Cruz knew well enough that his hours of freedom must be few. It was necessary to get on with the mission. It was arranged that he should meet a nephew of Mrs. Quezon in Quiapo Church. There would be means to escort him to still another contact, who would arrange the meeting with General Roxas.

On the streets again he found himself staring at the weary people, the smashed buildings, and the bomb-pocked pavements of this city he loved. There were the gashed trees along Taft Avenue, the ripped areas of the *Luneta* where gay Sunday crowds once strolled —the dirt, the stinks, the piled-up rubble everywhere. Yet this should all be routine to him by now.

Despite his resolution he halted as he crossed the plaza to the church and stared at a tiny Filipina girl whose abdomen was so distended it seemed that her thin pipestem legs could not support her. Malnutrition. He went on, then stopped again, horrified. At a more distant intersection he beheld a procession of ghost men— white men whose ribs showed through the rags and tatters of onetime khaki uniforms. They shuffled along like men in a nightmare procession. American prisoners of war. His feelings were in turmoil. He rushed into the church.

In another moment he was finding it hard to prevent open weeping as his hand gripped that of his old friend and they embraced. For a few moments all was forgotten except the reunion—whispered names, places, those gone.

Then Cruz felt the other's hand tighten on his arm and his almost inaudible whisper calling his attention to a man near the door who was watching them intently. Cruz did not turn his head, merely rotated his eyes. He saw the man approach Japanese police at the front door. Cruz and his companion went along the wall toward a side door.

He ducked through, and almost collided with a car parked close in front of the doorway. He started to skirt around the machine, when his eye fell upon the features of the Filipino behind the wheel. At the same moment the man saw him.

"*Por Dios!*" he exclaimed. "It is you—at last!" He thrust a door open. "Inside . . . *Pas . . . Pas* . . "

Cruz's companion pushed him from behind. The car jerked into motion. Cruz felt limp with reaction, for he had blundered into the very man who had been sent to make the contact, an old friend of the better days.

The next day Cruz shaved off his mustache; General Roxas had informed him that morning that the Manila police under enemy control were avidly seeking a one-time aide to the Presidente, an aide who wore a black mustache.

Perhaps the report of one Abduraman Ali had reached Manila more swiftly than Cruz had been able to come.

Danger faced him on every hand. And now whatever the threat to him there was one of similar dimensions to the Roxas household, for in those days no man could say who was his friend and who a collaborator. But Roxas, whose courage and integrity had girt him around against the worst or the most tempting the Japanese occupation could bring against him, now held daily conversations with Cruz on subjects of immense importance to both the military and the political fronts in Australia and Washington. Who among the one-time public figures had been forced to bend to Japanese will, or had apparently succumbed to blandishments? Who were true collaborators and who were merely playing the dangerous game the better secretly to serve their own people? What members of the old legislature had "gone over" and what ones could be depended upon to lead their constituents into anti-Japanese rallies when the time came to coordinate civil and military operations in the final showdown? Cruz made other contacts as well, sometimes with the very men whose apparent collaboration had made them the objects of suspicion, even hatred. There was still one more he wanted to see, the most controversial "collaborationist" of them all. Was he really so? Disguised as a vegetable peddler, Cruz went to see. But the man's garden was too full of Japanese soldiers to permit an entrance. There is little question but that Cruz's failure in that case saved his life.

From October 22 to the twenty-eighth he was truly a man who walked like a ghost in that city which was but a ghost of its former self. Then, packing a small bamboo trunk with papers concealed beneath cigars, handbags, and wooden shoes, he started south again. Pursuit was so close he could feel it, but somehow he felt, too, that he would not be there when the enemy sought to spring the trap. At Matnog in the Bicols it was that way, and again at Gigantangan Island when a typhoon blew a Japanese patrol boat into the town where he was hiding. Finally he made Negros and the *Narwhal*. When the Japanese came, once more he was not there.

In Brisbane General MacArthur's office had arranged for a decoration ceremony for Cruz. Three o'clock came and went. Cruz was not there. He had become so engrossed in writing his report that he had forgotten. That night, however, we had a celebration dinner at Lennon's Hotel for him. It was a honey. Cruz was there.

"Carabao Boy"

IT WAS AFTER HIS OWN RETURN TO AUSTRALIA IN November of 1943 that Villamor told the story of "Carabao Boy."

His agent had been well indoctrinated in the espionage dictum that boldness is often the best course. Finding it nearly impossible to locate undamaged studio space in the war-torn city of Manila suitable to his needs as an architect, he had decided to appeal to the *Kempei Tai*. The Manila unit of this dreaded Japanese version of military police occupied one of the most desirable buildings, and the Japanese had been making much propaganda about rebuilding Manila after "its destruction by the cowardly Americans." His appeal was rewarded. An office immediately across the corridor from the main police office was assigned to the cheerful Filipino whom the sentries seemed to find very amusing in his efforts to learn their language. The usual searches and checks eventually went by the board and often he stopped to bow and chat. Sometimes he heard things—such as pertained to great underground ammunition, bomb, and fuel dump developments at Nichols Field.

In due time a new water boy applied for a job there and got one. From the broad backs of their carabao, or water buffalo, these boys would serve the work details with thirst-quenching drafts. One of the most reliable, the most certain to be around, was the new boy, a good-humored chap who might have been a younger brother of the architect. The age of Orientals is notoriously deceiving and it had required only ragged short pants and a few other alterations to put him in a class with the others riding their slogging animals. He was ever obliging, especially to those who worked in the restricted areas of the underground dumps, the ventilating system installations, and communication alleys. Like any good Filipino, sometimes he curled up and slept on the platform-like back of his beast, while it stood shoulder high in a mud wallow.

A close observer might have noticed much activity for a sleeping lad, but he was careful that there never should be close observers while he was sketching out his "roughs."

Back in the office he made the "finals" with all the artistry of which he was capable. The sheets were sizable, however, and he was confronted with the problem of smuggling them out of the *Kempei Tai* building and rendezvousing with the "contact" who knew the time and place in the vicinity of Cape Santiago that the submarine would surface to receive them and carry them south

"The boldest course . . ."

He rolled the stiff sheets and stuck them into his hip pocket. When he had grinned and bowed to the sentry in the corridor, the roll had stuck out at an awkward angle. But there had been previous occasions when the architect had carried his drawings in that manner, and the sentries had become indifferent to unrolling them and examining them. Of course they might examine them this time, or there could be a new relay of guards who were unfamiliar with him, his habits and his papers. But it was an old team of sentries.

So uncanny were American bombs in finding their objectives in the Nichols Field installations when the day came for the avenging American pilots once more to thunder through the skies above Manila that the enemy command was constrained to release an official communique which asserted that the Americans had perfected a new aerial bomb which was attracted by concentrations of ammunition and fuel. (The same was to be said as a result of the work of secret observers reporting on installations around Davao on Mindanao.)

After his return to Australia in November 1943 Villamor remained in Brisbane only long enough to permit detailed debriefing of him by relays of G2 interrogators. It was a source of increasing puzzlement to him, however, that in certain high places there seemed to be an indifference to his findings and even glances down long noses that bespoke outright questioning of his sincerity. He had no time to go into these aspects owing to a request from President Quezon of the Philippines that he come to Washington and make a report in person.

The campaign was to continue in his absence, even to insinuations of discredit in official records. These were strenuously resented by myself, by the controller of AIB, by General Willoughby, and by others acquainted with the real facts. Nevertheless, the Commander in Chief elected not to become involved and it would not be until sixteen years later, in Washington, that in this connection it would be necessary for me to reshuffle an appointment at the Pentagon with Jess Villamor. The reason: he had been invited to the White House, where the President of the United States desired to make a personal apology to him for the injustice that had been done him. The President was warmly emphatic about his own evaluations of the little Filipino-American and the work he achieved with the "Planet" organization. In testimony to that, President Eisenhower presented Villamor with the official file of the case.

For Villamor it was one of the proudest days of his life. There could have been no finer vindication, no more solid tribute to all of those who had served in "Planet" in the beginning or in the endless ramifications of it throughout the troubled Philippines of that day.

Painstaking surveys by impersonal researchers after the war would reveal that his nets extended from the northernmost reaches of Luzon to southern Negros and from Panay on the west to Sorsogon on the east. Further, his agents were of a consistently high quality, reflecting, no doubt, the discrimination he exercised in choosing his schoolmates in younger days, and later in forming associations in the Philippine Army Air Corps PA AC. In fact, his work among the latter proved so cohesive that long after Villamor's GHQ critics had effected a degree of sabotage of his efforts, these loyal Filipino associates carried on with little or no official support, even increased both in strength and effectiveness, filing their reports with whatever organization would accept them. These were known as PAAC-AIB units. As stated, Villamor's original Station 4E7 on Negros was left under Andrews, to become one of the main links of the Seventh Military District. Through this station and "feeders" initiated by him not less than four hundred and sixty-nine messages of all types to GHQ are officially credited to Villamor, while more than twice that number eventually were credited to the Seventh District net control station under Lieutenant Colonel S. Abcede, which itself was established in consequence of his early efforts. Long after his return to Australia, and America, some of his agents and their radio units played key roles in reporting the movements of the Japanese naval units in the showdown Battle of Leyte Gulf in late 1944.

Chic Parsons, "The Artful Dodger"

EYE FIXATION ON SOME BRIGHT OBJECT—frequently a coin—coupled with a smooth, lulling flow of words have been trance-producing factors in the techniques of the successful hypnotist since the art was first practiced. Generally, however, the bright object has been fixed in one position. But Charles A. "Chic" Parsons was a nonconformist—even to the point of not conforming to one's idea of what a nonconformist might look like or how a nonconformist might act. He did not lay claim to being a hypnotist; he probably never considered it that way. Chic's coin was constantly in motion, up and down in front of one's eyes. Twelve times out of twelve he could do it. Twelve times it would come down tails if he called tails, and twelve times out of twelve it would come down heads if he called it that way. Accompanying the nonchalant, coinflipping motion would be the steady, low-pitched ratde of words, earnest but unemotional, convincing, making sense. Sometimes a ranking headquarters planner would basically alter his plans; sometimes a tactical commander would discover that he had agreed to commit a sizable portion of his command when originally he had entertained no such notion—or someone else would commit his loyalty, his immediate future, or his very life to the short, stocky man with the bland round face spinning that coin before him and talking.

Actually, Chic never flipped the coin over at all. True, he flipped it into the air for a couple of feet before it fell back into his other hand, and it certainly had the appearance of rotating over and over rapidly as it flipped.

The fact was that, as it left his hand, he merely —and very expertly—gave it an off-center snap with his thumb in such a way that it shot wobbling into the air and came back down same side up as before. Thus, if the "head" was uppermost when he launched the coin, "head" it was when it came to rest.

That was only one of the ways in which this amazing individual created illusions. From the standpoint of the Japanese command in the Philippines, he created illusions on such a grand scale that he became one of the most intensely hunted men in Southeast Asia. Although the reward for the head of Charles A. Parsons eventually reached a sum that was quite beyond the imagination of the simple Filipino guerrillas among whom he moved for prolonged periods during repeated penetrations of the Islands, not one of them ever caused him a moment's concern about betrayal. A hundred thousand pesos—fifty thousand dollars' gold—with all their power to twist the souls of men—could not buy Filipino loyalty and Filipino devotion to "Chicho."

In desperation, Radio Tokyo, on a masterful note of triumph, announced his capture and death. But in the Philippines the Japanese High Command kept right on hunting. That was in 1943 after the "Fifty" party had made its presence in Mindanao vividly felt.

The dose was a doubly bitter one for the Nipponese authorities to assimilate, for only a year before Parsons had been securely in their hands in the dank dungeons of old Fort Santiago in Manila—and they had let him go!

Born in the Tennessee hills, a neighbor of Jean Fairfield, who was to become Mrs. Douglas MacArthur, Parsons entered the Philippines the first time in 1921 as a member of a freighter's crew. He was lured there by the tug of an imagination fired by the tales of two uncles who had preceded him years before. Trained in stenography at home, Chic became secretary to Governor General Leonard Wood and in that capacity traveled the intricate pattern of the archipelago for three years, learning the land, learning the people. He responded to the chromatic loveliness of the former and to the warm, honest simplicity of the latter. It was mutual. Chic knew he had found his corner and his people. He never changed. It was that direct, that final, when he met Katraushka Jurika in Mindanao.

For Chic and "Katsy" it became hand-in-hand progress through the eventful days of his emergence as a business leader with the Luzon Stevedoring Company—and in the subsequent days of disaster when the Japanese military machine thundered in and crushed organized resistance in the Philippines.

Together with their three young children they could have sought safety in time. But this was his land and these were his people. So he stayed, working like ten men to move all possible equipment and all possible supplies from the Luzon company's great wharves in Manila to trucks and ships waiting to make the increasingly perilous runs to Bataan and Corregidor. MacArthur had declared Manila an open city to save it from destruction, and the enemy was entering from the south.

Then came New Year's Eve, 1941.

Chic and Katsy entered the Army and Navy Club. Only a few short weeks ago it had resounded to the gay pre-Christmas parties of American service personnel. Now—silence. They stared at the deserted tables. On the dais was a Filipino band. The round eyes of the silent musicians followed them as they moved slowly forward. Only one other couple was there. From the walls a flickering red light was reflected. Chic and Katsy knew where it came from: with their own hands they had set the torch to the last of the stores on the great wharves so the enemy might not use them.

The hour struck twelve. For a long moment there was no sound. Then, doubtfully, the Filipino band began. It was "Auld Lang Syne." Chic and Katsy stared at each other. Then, wordlessly, they pulled each other close and danced, and tried not to think—or smell the soot and gasoline on his uniform and her party dress.

On New Year's Eve a year later, Chic was enjoying another dance with Katsy—this time in Washington. In his pocket were his orders directing him to report to Brisbane immediately after New Year's. They recalled "Auld Lang Syne"—and what had followed.

Chic had burned his naval uniform and papers. Together they had packed what they believed the enemy might allow them to have in concentration camp. And then Chic had an inspiration. After all, American citizen or no, he also was the Panamanian "honorary consul."

He had served in that capacity for some years until a "career" man might come from Panama. He had the seals, he had the stamps. Why not have the diplomatic immunity of a neutral, too? In his heart he knew why—only "career" consuls were entitled to that kind of immunity. But would the enemy commander know that?

There was fast action around the Parsons menage. A big sign appeared on the gate. A Panamanian flag was recovered from someplace.

They were just in time. The first Japanese patrols were entering the city. In a short time the military government was installed and it was combing through every district with devastating thoroughness. It came their turn. Chic brazened it out. It looked as if he might be winning. The Japanese authorities decided to recognize his "status." A certain amount of freedom for him and the family resulted– until one day he was suddenly ordered to Fort Santiago in the old Walled City area from which few returned. He was tossed into one of the underground dungeons of that gloomy, wet old Spanish keep.

But Chic had not lost heart. Instead, he became increasingly vehement in his demands that his "status" be respected. Eventually he managed to make an outside contact, one Helge Jansen, another "honorary" consul whose status the enemy had accepted. It was embarrassing for the Japanese to explain how they recognized one, but not the man who represented the neutral Republic of Panama. Jansen sent secret messages through Sweden to Panama. Panama protested.

There ensued a sudden trip for the whole Parsons family by air to Hong Kong and eventual repatriation via the Gripsholm.

AIB learned of it through Peter Grimm, Chic's Luzon Stevedoring Company boss, who had become a colonel in the Army Transportation Corps in Brisbane. Immediate request was made to GHQ for Chic. General MacArthur already had taken action to bring him over "soonest."

This was early 1942, and Villamor's key station in Negros was beginning to pass traffic of a kind that was proving very helpful to us in the preparation of "Fifty" Party. The information was concerned primarily with logistical needs, since the guerrillas on Mindanao, mostly under the leadership of former mining engineer Wendell W. Fertig, were sufficiently entrenched in some areas to give reasonable assurance that supplies could be landed safely. The mission of "Fifty" had been altered from one of expanding clandestine activities of the "Planet" type to giving all possible logistical support to Fertig in the faint hope that Mindanao might be preserved for use as a base for ultimate reoccupation operations. There would be materiel and money for Peralta on Panay as well.

Amid a welter of activity involving the soldering of tons of equipment into waterproof tins at the AIB supply depot in Brisbane, Parsons saw a figure that made his brown eyes light up with joy. The greeting between Chic and Charlie Smith was almost casual. But that was not the way either of them felt. They were old acquaintances from the prewar Philippines and both confessed that neither had expected to see the other alive again. Smith surveyed the gold braid of Chic's smart naval uniform.

"I'm in the army, Chic, for one thing because I know I'll never be a sailor."

Charlie Smith was a mining engineer from the island of Masbate in the middle Philippines. Together with his partner, the tall, rangy Jordan Hamner, he had duplicated the feat of a few other intrepid amateur navigators and had succeeded in spanning the tremendous distance to northern Australia in a small boat. Burned black by the sun and emaciated, they had made landfall near Darwin and, as mentioned, were brought to Brisbane by plane. Almost at once they volunteered to augment the already considerable store of information they had given G2 about conditions in the Islands by volunteering to go back in as spies. It was agreed. Charlie would lead "Fifty," Hamner would go in later and try to extend communications and observations along the Sulu Archipelago toward Borneo. Each party would include certain Moro tribesmen who had formed the crew of still another amazing band to sail their fragile craft through weather and Japanese alike from the Philippines to Australia.

Parsons looked around him. There were radio transmitters, receivers, charging units, batteries, t.ools, wire, and spares. There were sidearms and ammunition and grenades. In another area were bundles with red crosses on them—drugs, surgical kits, vaccines.

Being chucked into every niche and cranny of the sealed tins were copies of newsmagazines, newspapers, bars of soap, sewing kits, cartons and cartons of cigarettes, and individual packages where the cartons were too big to go in. Chocolate bars, socks, shirts, underwear. There were lieutenant colonel's insignia to go with the official commissions establishing Fertig on Mindanao and Peralta on Panay as the GHQ-appointed commanders of the Tenth and Sixth Military districts of the Philippines respectively.

"There's two things wrong, Chic," said Smith in that soft, almost lisping way of his. "First, there can never be enough of this stuff—and, second, you are not heading this penetration. How about it?" The coin came out and went flipping into the air. "Heads you go," murmured Chic. "Heads it is."

"No, you don't. Let's put it up to the bosses. You're the man, Chic."

GHQ considered carefully. It was certain death for Parsons if he were retaken by the Japanese, and a terrible death. The coin spun and the soft patter went on in the Commander in Chief's office. Two days later the decision came. Chic Parsons would head "Fifty."

"Swell, Chic," was all Charlie said as he grasped Parsons' hand. Training and preparations went into high gear—an eighteen-hour schedule. Everywhere the Parsons' kind of foresight was registering, things that proved important out of all proportion to their size. There was, for instance, his request for a sealed tin of buckwheat flour.

"But why?" he was asked. "There are fifteen million Filipinos up there, all hungry. What good can one tin do?"

"Fifteen million, right—and most of them Catholic. This is for Communion biscuits. Spiritual food. Gotta get wine, too."

Now there was only one grave deficiency: somewhere between Washington and Brisbane there was a shipment of money, Philippine peso notes, needed by Fertig and Peralta as urgently as food, medicine, and arms if they were to live to fight another day. Washington reported that the fortune in notes had been shipped by special air. What was not known was that to mislead the curious the consignment had been labeled "Finance Forms." A determined delegation of us from AIB drove through a night black with rain to check with Amberley Field, thirty miles or so from Brisbane. Sodden on the flooded hard stand where they had been unceremoniously dumped the previous night by a weary air crew before taxiing off for hot coffee and sandwiches were the abandoned cases of "Finance Forms." Fifteen hours of counting ensued. Then the soldering irons. And so it was that, early in March 1943, at the same time Malcolm Wright and Peter Figgis were standing off the dark New Britain coast aboard *Greenling*, and readying their kayaks, the U.S.S. *Tambor* was boring her way far to the westward with "Fifty" bound for Pagadian Bay on the south coast of Mindanao. Aboard were Chic, Charlie Smith, and three Moros, together with seven tons of equipment and miscellaneous cargo. There was complete compatibility between Chic and *Tambor's* skipper, Lieutenant Commander S. H. Ambruster. It was a dangerous combination for the Japanese.

In the dry language of the Tambor's log, target area was reached on schedule:

4 March 1943:
    1850 Hours—Surfaced and proceeded toward Pagadian Bay.
    2000 Hours—Removed wherry from skids and lashed it on deck to avoid noisy operations close to beach.

5 March 1943:
    0445 Hours—Arrived four miles southeast of Labangan. Visibility poor during night; practically all navigation by S. J. Radar.

0517 Hours—Lieutenant Commander Parsons and two natives left in wherry. Submerged and patrolled area.

1637 Hours—Sighted wherry flying white flag and proceeded to close it.

There had been enough exchange of messages between Mindanao and GHQ to assure that Pagadian Bay this time ought to be safe for landings. But who could tell? Chic had drawn a long sigh as behind him he saw white foam and bubbles dissolve into the engulfing darkness; the *Tambor* had gone back down to wait silently on the bottom. Ahead was only the dark smudge of the land.

The two Moros paddled quietly enough, but there was no enthusiasm. Chic whispered quick words in Spanish to put energy into their strokes. The shore approached them and seemed slowly to engulf them.

They were almost in.

Suddenly there were red and yellow spurts in the shadows. Bullets whined as they ripped against the water and ricocheted.

The Moros froze at their paddles. In mingled English and Spanish Parsons tried to break their paralysis. He knew that as they lost way, even a poor marksman would be able to hit a sitting target. Then came two more shots. Chic slapped the nearest paddleman smartly on the back and yelled in Spanish. The man began to paddle hard.

They went straight onto the beach. With a silent prayer Chic slowly and carefully got himself out of the wherry and stood. A rifle suddenly poked out of a mangrove thicket and steadied exactly between his eyes. Now *he* froze.

How could he somehow make the rifleman reveal his military status before it was too late? Then he saw the man's foot appear beneath the leaves. It was bare. The Japanese always wore shoes. His voice was husky with relief.

"I come from General MacArthur. I bring letters, supplies— cigarettes. See!"

The guerrilla came out to see—and what he saw sent him shouting back into the jungle.

The party was on! Such a vociferous, pathetically eager welcome Chic had not imagined in his fondest moments of ultra-optimism. He had landed in an outpost zone of Fertig's own waiting guerrillas. Hundreds of willing hands reached to carry his little armful of supplies—and him, too. He told them when he could that the rest of the stuff was still on board the submarine. How could it be gotten off?

The local leader beamed. He would show him how. Just "follow me only."

Chic stopped, amazed. Anchored in an inlet was a small lighter, complete with crane and winch. "... a present from the Japanese." There was an excellent launch, too.

As per prearrangement Chic stood out in the wherry first, showing the proper signals. The barely surfacing eye of the submarine fixed upon him. The visibility had not been good, and now a gray rain obliterated the shore line. Ambruster considered. It could be a trap.

They surfaced. All defenses were manned. It was Ambruster's turn to gasp. There was not only Chic, but a miniature navy, complete with unloading lighter!

With a now well-armed patrol, Chic, Charlie Smith, and hilarious guerrillas started for Fertig's hidden inland headquarters. No one slept. But for once everyone smoked—and talked, talked, talked.

Then came the inspections as per GHQ instructions. Days went by. Then down over the miles Chic sent his first long cipher message in his own key; and with it a greeting from provincial officials for President Quezon in Washington.

The time came for the separation of Parsons and Smith. In keeping with a master plan prepared by AIB and approved by GHQ, Coast Watcher stations were to be established without delay in the Surigao Straits region of northeast Mindanao—where fate was to decree that one of the greatest naval battles of all time would be fought in October of 1944—while a second station was to go into the Davao area in the southeast. From a military point of view the Surigao Straits ranked with the San Bernardino Straits farther north and with Manila Bay as the three most sensitive sea approaches in the Philippine Islands. Chic would go to the Surigao area. Charlie would go to Davao. (Eventually Katsy's brother, Tom Jurika, would do yeoman work here.)

"*Mabuhay*, keed," grinned Chic, squeezing Smith's hand. He watched him disappear into the hill country with his picked assortment of Moros and Fertig guerrillas. Charlie's carriers were supporting cases containing little ATR4A transceivers, as well as one Dutch special set for longer communication. This one would be safely hidden in the hills, the little pack set could move up close—theoretically, anyway.

Grave, intense, efficient Lieutenant Colonel Fertig turned to Parsons and pointed to his prize possession, a sixty-foot Diesel motor launch, Japanese-made and, until recently, Japanese-owned. Chic knew well how loath the hard-pressed commander was to turn over the *Nara Maru* to him, but it was the only way if he was to do his assigned job. Traveling by night and hiding by day should enable him to make it.

"She's ready," said Fertig. "Run for Medina. That's Ernest McClish's headquarters on the north coast. I've sent him a message. He's one of my best district commanders. A bit impulsive in his desire to liquidate Japs, perhaps, but..."

McClish, a one-time businessman of Manila, was well known to Chic. Their reunion was a glad one. Chic was to refuel. But with what?

McClish's service company proved that it could provide an excellent substitute for Diesel fuel for the ship's motors, or whisky for the captain—all in the same liquid. It was accomplished by taking the flower of the palm tree, withdrawing the sap, fermenting it, and distilling off the alcohol in stills made of battered five-gallon tins and rubber tubing carried under the surface of running streams to provide the condensing media. Amazingly efficient, it would propel a boat or a man according to the control—or lack of it— invested.

McClish also armed the *Nara Maru* with the only .50-caliber gun in the area—one of the old Air Corps guns salvaged from a smashed B-17 of the original Nineteenth Bomb Squadron. Equipped with a piece of rubber tubing from a discarded auto tire as a recoil spring, it would fire as a machine gun. Without the tubing, it was a singleshooter.

Eluding Japanese patrol boats by hiding in the daytime, they reached the country around a big lumber mill in the Agusan River area. Chic was playing a lone game at the moment. He did not seek the hardy "king" of that part of the country, Buck Walter. But some of Buck's one-time workmen gave him news that made him thoughtful: the Japanese were in full control of the Surigao area, at least on the Mindanao side. There would be no possibility of operating a station there. In fact, added his informants wryly, enemy patrol boats based on Surigao were out looking for him at that very moment.

Chic was not surprised at the news—any of it. The Japanese knew the Straits to be a vital seaway and were watching it. Chic wanted this station to be on Mindanao to simplify land communications with Fertig and to insure its being within Fertig's official control. The only alternative was to move either across the Straits of Leyte, or to some intermediate island and set up a watcher station to watch the Jap watcher stations—and everything else.

There were two islands off Surigao. One was Dinagat. "Likely hot," considered Chic. He selected a smaller one to the west, Panaon. Its proximity to Leyte would facilitate escape to that large island if the Mindanao side became untenable. Besides, there was work to do on Leyte, too.

The dash was made after dark. The *Nara Maru* was not the only Japanese-made craft aboard in the Straits that night. Chic never knew whether they were mistaken by one enemy patrol boat for one of its own kind, but at one electric moment they went unchallenged within hailing distance of a prowling enemy searcher. At dawn they slipped into a well-hidden anchorage—straight into the arms of a guerrilla force whose faces were savagely streaked with yellow.

"War paint?" queried Chic, breathing easier after he had managed to establish his identity.

The explanation was that the guerrillas had been particularly bothered with tropical ulcers. They had snared a big Japanese mine on the loose in the Straits and had dragged it home. The detonators had been removed and the picric-acid powder recovered to treat the ulcers.

Chic knew that men of that caliber would not let him down. He left them in charge of a beautifully concealed little station on the east coast. But Panaon offered little maneuver room. To Truman ("Ernest") Hemingway, one of Fertig's best men, Parsons did not mince anything.

"This is one of the most important 'eyes' GHQ can have," he said prophetically. "And you're it. The enemy is just over that hill there, and they're almost sure to get closer."

Hemingway made no boasts. Chic knew his man; far-off Brisbane was to know him, too. Through increasingly critical days and nights, Station "TUF" would be whispering into Fertig's control station with priceless ship-movement information, even when the enemy was combing the tiny island itself for the station that was proving so costly to them. Captured Japanese intelligence documents later revealed that the enemy had determined to clean out the tiny hornet's nest and had sent a strong punitive expedition against guerrilla forces in southern Leyte and Panaon. The Japanese patrols prodded every bush with bayonets. Station "TUF" sent a brief warning to Fertig, slipped out under the very rifle muzzles of the Japanese, and skittled across to Dinagat. Until the end of the war "TUF" continued to pump in its precise ship "fixes" for American submarines that hovered within calling distance beneath the waters of Surigao Straits—the graveyard of so many of the Emperor's vessels.

After establishing Hemingway, Chic moved cautiously to southern Leyte, where guerrillas battled not only the Japanese but just as savagely fought each other for local control. Somewhere was Colonel Ruperto K. Kangleon, former district commander of the Philippine Army for Samar and Leyte. It was Chic's hope to find him, and that the venerable and thoroughly respected Kangleon would agree to serve as guerrilla coordinator for the important island area and put an end to the senseless warfare. Chic began his search. The grapevine had served; guerrillas on the extreme southern end of Leyte were enthusiastically friendly.

He had requested the local guerrilla commanding officer to warn those farther up of his coming.

"Yes, it would be advisable," the other had replied. "I have not yet known of a Japanese launch to get through, or anyone aboard to—ah, escape. And your *Nara* has a nice Rising Sun painted on both sides and on the deck. Good bull's-eyes!"

Good bull's-eyes they proved, despite the added precaution of an American flag Chic had caused to be hoisted before they moved closer to the richly green, quiet shores. As they turned toward a barrio where Chic had reason to believe the commander of that area would be waiting for him, they were met by a volley of rifle shots.

Chic scrambled on hands and knees to reach the wheel. In his mind was one compelling thought: they must get in closer so the hidden riflemen could see the American flag. He said later it had not occurred to him that the fire might have been coming from Japanese who would not have been particularly panicked by the sight of the flag.

Someone yelled that the fuel drums had been shot. Chic bellowed an order for all to lie still. He raised his head a fraction, just in time to see the man who had cried out leap to his feet and run to the forepeak, shouting and gesticulating to the shore. It was no use to order him back. Every moment Chic expected him to drop with a slug through him. Instead, the firing on the shore ceased. The Filipino's crazy appeal had worked.

For a moment no one moved. Then several men jumped to stuff emergency plugs in the spouting fuel drums.

No one had been hit, though the little vessel bore some fresh scars. The turn in toward port had given them some protection and had reduced the size of the target.

An apologetic lieutenant explained: The message he had received was missing in one vital word, "American." He had been warned to watch for "a launch" in the Straits that afternoon and had prepared two reception committees. If this one had failed, three quarters of his riflemen were concentrated at a certain narrow point farther on. After all, the only launches they had seen since the fall of the Islands had been Japanese.

Parsons resumed his survey. Little time was required to confirm messages sent down previously by Villamor's 4E7 that conditions on Leyte were worse than chaotic. As Chic phrased it there were "at least six first-class wars going on not counting the official one." Two were being waged between a former yeoman in the United States Navy and a mining engineer. Then there was a private one involving Lieutenant I. David Richardson of the Navy. Richardson had plenty of what Chic needed in the way of leaders and he was, as Chic put it, "a fighting fool who knew when it was best not to fight." There were some lesser private fights, too. Unification had to be. In Chic's book, Kangleon was the man to achieve it—if he could be prevailed upon. Young Richardson should be his field boss.

Southern Leyte guerrillas knew of the "Soap Factory." To anyone else, that was exactly what it was—a place where an old man and his big, good-natured wife made soap and sold it for forty centavos a chunk. After Parsons' historic visit with the leathery-skinned old patriot, the "Soap Factory" had another meaning, one with more than cleanliness in it—one with grim hope in it.

It took all of the earnest persuasiveness Chic ever had to convince the war-weary, physically unwell Colonel Kangleon. Automatically Chic began to flip the coin. Then he talked. First he recalled for him the desperate plight of the guerrillas—without medicine, possessing only junk radio outfits that could not power the ether enough to reach more than a few score miles, homemade "paltics" for guns, Chinese firecrackers for powder to make homemade bullets.

There would be no overnight conversion to military strength, he made it plain, but a start had been promised by MacArthur, and Admiral Kinkaid was making available the submarines to carry the stuff in—provided only they could be assured that it would be worth the effort, that there would be a responsible man on the Philippines end to see that it was used for the good of the cause. He used the phrase then that he was to use often: "After all, a guerrilla movement is only as good as its leader."

Two days later our AIB decoders in Brisbane were extracting sense out of the five-letter cipher messages that were coming down from the north in relays—Leyte to Fertig, Fertig to KAZ, and KAZ to Brisbane.

KANGLEON HAS AGREED . . .

It was an agreement that was to withstand the combined attrition of jungle, of hates, jealousies, frustration, sickness, and sudden enemy raids, one of which was so sudden that the Japanese even captured Kangleon's new transmitter sent in from Mindanao, though they failed to get Kangleon. It was an agreement that would be confirmed wordlessly in a steel-like grip of hands in October of 1944—MacArthur and the proud old patriot whose courage and tough determination never had failed him after Chic's visit.

Parsons always traveled unarmed. It was his agility, he would say, that kept him out of trouble. AIB knew that his agility was twofold—mental as well as physical. Now he had another assignment that was going to test the latter.

On reading Chic's reports, the Commander in Chief had expressed his concern about a certain guerrilla leader, one "Brigadier General" Salipada Pendatum. It was feared that not only would Pendatum insist on independent operation in Mindanao, but that his utter determination to engage the enemy on every occasion would bring down crushing Japanese retaliation. Such a campaign by the enemy could not only wipe out Pendatum but threaten all of Mindanao.

Pendatum, a one-time lawyer and influential advisor to the governor of Mindanao, was brainy, proud, and unquestionably as good an organizer as he was man of action. That he was ingenious and ruthlessly determined had been proved when, upon laying siege to a strongly built structure in which the Japanese were laughing at his attempts to dislodge them with small-arms fire, Pendatum caused a water buffalo to be saddled with a hundred-pound aerial bomb, "aimed" at the building, and "fired" by means of burning cloth under his tail. Terrified and full of pain, the animal charged the building. The bomb exploded, demolishing a wall. The victory was complete.

As usual, Chic went unarmed to tell Pendatum that he must come under Lieutenant Colonel Fertig as a major. Fertig reported that he was more worried than if Chic had been bound for Manila itself.

"This will be better than a gun," Chic had said, grinning and patting a bulging knapsack on his shoulders. "Maybe it will get the self-appointed star off his collar."

In the knapsack was his proof that he came from GHQ. Probably no stranger credentials had ever been offered: cigarettes, candies, anti-malaria tablets, and several copies of *Time* magazine showing a mangled Japanese warship at Midway. Pendatum was not gullible, neither was he a fool. But this was proof. He took the major's bronze leaves.

THE PHILIPPINES					153

Originally it had been planned that Parsons would remain in the lower Philippines from thirty to sixty days. But so valuable was his work among the guerrillas (a task which Villamor could not do and at the same time organize the clandestine nets independent of guerrillas, his original mission) and so helpful were his reports to the Commander in Chief that weeks ran into months and still he stayed on.

In the interim there had been changes in the Philippine section of AIB. The flow of military, political, and economic information from the north was assuming impressive proportions; obviously to evaluate much of it required an individual whose knowledge of the Islands and the people that made up its social and political shadings far exceeded my own. Such a man was Colonel (later Major General) Courtney A. Whitney, who had resided and practiced law in Manila for more than twenty years. In mid-1943 he arrived from America and was assigned to take over the Philippine section. At his request and by direction from the "front office," I remained as his advisor on technical operations and consultant on policy; this would be in addition to my other duties as deputy controller and finance officer of the Bureau. Perhaps it ought to be mentioned here, too, that shortly thereafter I failed properly to evaluate and interpret the Willoughbian eyebrow and found myself "ranked down" to the level of "assistant deputy controller" under one Colonel Colin S. Myers, a grizzled old campaigner who did not know much of clandestine operations but did know army regulations. I learned a lot about them, too, after that. We collided often, but later served amicably and productively together in Java and Japan on important jobs. (In 1951, in Tokyo, I would be restored to the position of eyebrow interpreter *dai ichiban*—or Number 1— but that would be after the deactivation of the Bureau.)

It was shortly after Whitney had taken over that we received a startling message from Fertig. Whitney referred it to me. The text spoke of the white men who claimed to have escaped from an enemy prison farm near Davao after having been taken on Bataan. They had, they said, survived in the "Bataan Death March." Their stories, warned Fertig, would shock all of America—if they were true.

One name caught my eye: "Captain Ed Dyess." If the man claiming that name was genuine, he would know the answers to some questions I could send, because Ed had been the "squadron" leader of what was left of our Air Corps on Bataan—at that time, six P-40's. I was directed to make up the questions. In consequence of their exactly proper answers, Chic Parsons got his next assignment: he would escort as many of the men as were fit to be moved to Australia via a submarine that would come close to the place where the wretched men were recuperating. That place, Chic radioed back, could be a certain north coast point. Task Force 77 radioed to one of their "fish" and gave her the coordinates. Quietly the U.S.S. *Thresher* swung her bow around and got under way. Other messages went north that night in the personal code of "Q-10," Parsons, confirming the exact coordinates for the rendezvous. It was necessary for Fertig's operator to relay the message to "Q-10," who was hidden on the northeast coast. Atmospheric conditions were not too good, repeats were necessary.

The repeats that night were appreciated by certain Japanese wireless operators manipulating carefully calibrated dials mounted on powerful receivers. One set was in a vessel drifting quietly out in the Mindanao Sea just under the island of Bohol. A second set was on a craft slowly drifting in the Surigao Sea just south of Leyte. These ships were Japanese Radio Direction Finders—"RDF's." There would be quick calibrations followed by a rapid exchange of signals between the ships as the clever triangulation operators charted the next location of a troublesome guerrilla station. Japanese Naval Air Operations would be glad to hear of this "fix."

Two days later the enemy came in.

"Many *Hapons*... Many *Hapons!*" had cried the runners of Fertig's guerrilla army. Nimble hands disassembled small radio sets. In a trice there were no telltale aerials—the wires had collapsed into trees; and the little nipa shacks looked bare and disused by the time the advance reconnaissance parties arrived.

It was a near thing. Three times Chic's policy of sending ahead of him a simple Filipino who could be an innocuous tiller of *camotes* or peddler of bananas had saved him from heavily armed patrols. He did his "swan dive" into the jungle until the immediate danger, unobtrusively signaled by his "farmer," abated. But it was soon clear to him that this was no ordinary patrol. Reports came in from outposts everywhere of more enemy coming ashore. There had been short shrift with anyone who failed to cooperate. Chic sensed that either there had been a compromise, or a dreadfully potent coincidence. He dared not even try to break out the radio again—not then.

Anxious days later, back in Brisbane, a wordy message in the Q-10 (Parsons) code finally came through.

ARRIVE HQ TODAY FROM NORTH. WARN SUB COMMANDER
SITES ONE AND TWO FOR MEETING NORTH COAST IN

HANDS ENEMY SITE THREE NOT FEASIBLE. RECOMMEND MEETING ONE HOUR BEFORE SUNSET NINTH VICINITY NORTHEAST COAST OLUTANGA ISLAND FROM SAME LAUNCH THAT MET SUB. IF POSSIBLE DATE THIS MEETING SHOULD BE ARRANGED SPECIFICALLY TO AVOID ENDANGERING SECURITY. AREA SECURE AT PRESENT NO AIR OR SEA PATROLS. PORTABLE TRANSMITTER MALANGAS WILL WARN IF SUDDEN CHANGES. MORE TOMORROW.

Only immediate action might save the submarine from a disastrous ambush. There was not even time in Brisbane for us to take a staff car to Naval Operations. Instead, the open phone line was risked to say that "the threshing machine definitely must not enter the new field to work the harvest. The field boss says she'll bog down in yellow mud with disastrous results. Been raining in that country pretty bad."

Navy's fervent "Roger" clipped off.

From the tall towers in Perth the warning went north. To the intense relief of everyone the *Thresher*'s own "Roger" came back.

Now began one of the most trying forced marches Chic ever was to experience. Deprived of the relatively accessible site to the north, it would be necessary for the party to negotiate treacherous trails over some of the most punishing country in the Philippines to reach the area where Chic had made his original landing in March. And all this to be imposed upon a little company which collectively was in such poor shape that only three of the seven original escapees could be considered for the trip. They would be Dyess, Lieutenant Colonel Stephen Mellnik, and Lieutenant Commander Melvin H.

McCoy. With them also would be Charlie Smith. He had accomplished his daring job of establishing a watcher position above the city of Davao with its important port and signals from his little ATR4A had been feeding in excellent grist for the mills of Naval Intelligence, who gave it to Operations, who gave it to submarine commanders. Chic would be glad to have Charlie. Brisbane directed him to accompany the Parsons trek. It would be necessary for them to move out at once if they were to make the rendezvous at all.

During the morning of the first day they had made fair progress. But as the heat increased, the pace slowed and finally stopped. While they were halted there in a defile with the inhospitable mountains rising higher and higher before them, they caught the hint of smoke in the hot air. One of Chic's scouts slipped back from his forward position and signaled them to follow him. They stopped on the edge of a steep gulch at the bottom of which rushed a deep, turgid stream. Sagging into the swirling waters were the charred remains of bridge timbers, still smoking.

Chic knew that they were not far from friends. He was sure guerrillas had done this—and if they had, that meant only one thing: the enemy was in the area in some force. They had done it to protect the barrio.

Dyess muttered that they'd have to swim for it. Chic regarded his wasted body and knew that would be out of the question. Anyway, there were crocodiles, lots of them.

They would have to build a raft.

It was bitter work. The current sucked at their legs and pulled their feet from under them. The timbers from the bridge ruins would have invited the service of a crane hoist instead of these weary, famished bodies. Three o'clock came.

Chic called a ten-minute halt. Then he forced his own aching body to resume. A longer rest would be fatal; they would not be able to get going again.

By late afternoon the stream of the crocodiles had been navigated at last. Drearily they tramped the jungle trail, but their sagging bodies were sustained by the knowledge that just beyond should be the barrio they sought. A friendly barrio, said their guide. Rest, food.

Suddenly Dyess, in the lead, halted. Before he could signal, the others had plowed into him, walking blindly as they were.

"Japs," Dyess whispered. "Right there in the yard taking a bath."

For a moment they stared. Then Chic gave the signal for them to fade into the jungle. But some movement had caught the attention of the alert enemy. There was a sudden cry, and the soldiers clawed for their clothes and their weapons.

In a trice the party melted from the trail. Fear put strength into their tired muscles as they crawled away through the jungle.

Night closed down on them.

Still they went on, climbing steadily as they knew they must if they were to find the trail again.

At last they stumbled into a sudden clearing.

Mellnik's voice was a croak.

"Com . . . Lookit, guys . . . Com!"

McCoy seized an ear, ripped away the husk, and sank his teeth into the raw kernels. For a moment the others watched. Then savagely these famished men stuffed the green corn into their mouths and blinked back salty tears at the sheer ecstatic taste of it.

Their luck had changed. Gladly, humbly, the native shared his poor shelter with them. Instantly they were asleep, secure, for the simple Mindanaoan steadfastly refused to close an eye.

At dawn they were on the trail again, always upward. Would it never cease?

Then the guide halted, pointing to the earth. Chic and Charlie struggled to him.

In dismay they recognized the fresh imprint of a Japanese soldier's "tabby shoe," a soft type with a separate "toe" for the big toe. It also pointed *up* the trail.

So while they had rested the enemy had passed them and now were "pursuing" them up ahead. They held a conference. Then they pushed on, for that way, at least, they knew where the enemy was.

Slipping, sliding, and fighting always, they mounted.

Suddenly the guide signaled a warning. Peering through the little clearing he had made, they saw what he saw.

The enemy patrol was comfortably spread out in a sheltered hollow, and—they were eating!

Dyess cursed. He pulled his gun. But Parsons reasoned with him; it would be a hopeless attack.

Charlie Smith was reconnoitering. He came back to suggest that they find a way around the enemy and go on, for it was his bet that soon the Japanese would suspect that they might have outrun their quarry and turn backward. This was done. Later they began to drop down into lower, hotter country. Twice the guide halted, seemingly for a rest. Smith observed him. He knew Filipinos. Was the man lost? He faced him and put the question he hated to ask.

Yes, he was lost.

For a moment they were stunned. Then McCoy began to speak. There was a flat pitch to his words that made Chic jump up and put an arm on his shoulder, and say reassuring things. McCoy became silent. Chic and Smith conferred. Charlie pulled a little compass from his pocket. He could do no worse, he said; from now on he would be "guide." It was agreed. It was then that they discovered that they had only part of one canteen of water left.

There was no alternative: they would have to trust stream water. Later they dropped on their bellies beside a small stream and filled themselves and their empty canteens.

The frail needle of Charlie's compass tremblingly waved them on through hellish, endless days, steam-infused and miasmic. The nights brought no rest—only the dawn, and the jungle again, crowding, implacable. ...

At times the ex-prisoners seemed to lose all sense of time. Parsons wondered if there would be a tomorrow for them? Their strength was going fast. Yet back of them all certainly were those well-fed stocky Japanese. There would be no quarter, asked or given. Ahead, someplace, would be the submarine. It had to be soon.

His grim musings had been suddenly interrupted by sounds of someone ahead. They pulled each other into the undergrowth. Could the enemy possibly be ahead of them, another patrol? Then Chic jumped as beside him Charlie Smith called out loudly:

"Guerrillas! Hey, Captain Medina! It's *friends!* From Fertig! ... Christ!"

The reaction had been terrific. Tears streamed down their faces unashamed. There was joyful sharing of meager rations by the little guerrilla band.

The guerrilla leader jerked his head back along the trail and promised that he would do things to discourage the Japanese from overconfidence.

They went on, momentarily encouraged and refreshed. Medina's ragged fighters had disappeared to the rear. (They did not hear the firing of the ambush. Some enemy got away, it was known later. Medina and his band would be pushed high into the hills, where they all but starved.)

Another day and night went by. And now it was actually the sixth. They dared not lose more time. Yet the ex-prisoners were at the end of their strength. They appeared to be dazed. Sometimes they would wander away. And then, when it seemed that even the comparatively fit men could endure no more, they were confronted with another great hulk of a mountain. Beyond words now, they started. Up and up.

Somehow they breasted the summit. There they halted, words dead in their throats.

Below them was the bay.

Behind guarded doors at AIB headquarters in Brisbane the interrogations were pushed as rapidly as the strength of the pale, drawn-featured escapees would permit. There were other interruptions as well, because the stories were so horrifying that the stenographers could take it for only twenty minutes at a time. For these were the first participant narrations to come out of the Philippines of the Bataan Death March and of the unbelievable prison camp life following it. On orders of the Commander in Chief, the word-for-word record was sent to him immediately for special air dispatch to Washington, because the United States Government had decided on a nationwide expose fully to acquaint the people with the nature of the military enemy that was ravishing the Pacific world.

As indicated previously, Chic's return to Brisbane in July 1943 coincided with the extensive expansion of activities in the Philippine section of the Bureau. It has also been stated that before his return other penetration parties had been organized to follow "Planet" in, such as "Peleven" and "Peleven Relief." There was another, known as "Tenwest," led by Charlie Smith's boat partner, Jordan Hamner. By the time of Chic's return, "Tenwest" had gone and had insinuated itself into one of the most unenviable positions in the whole area—that of the Sulu Archipelago, a string of islands extending in a generally westward direction until they almost touch the northeastern part of Borneo. This screen, frequently referred to as the Tawitawi, had to be watched; enemy movements from the western Philippines, or even from Singapore, headed southward toward our island-hopping troops on New Guinea, in all probability would set their course through this line of the Sulus. It was a tactically delicate area; it was also notoriously a hostile one.

Parsons' reports encouraged GHQ planners in their hopes that if every assistance within our admittedly limited potential were extended to Fertig, and if a military policy of non-offensiveness toward the enemy were strictly observed on Mindanao, possibly that island might be preserved long enough to enable the Allies one day to swoop upon it with a force sufficient to hold it as a base for a big northward campaign. The problem, then, was to increase the flow of necessities to Fertig and, if we could, to help the others stay alive until "the day."

The consensus of numerous conferences was that attempts to do this by aircraft would prove unfeasible; even if we had the aircraft, their presence would only alert the enemy of GHQ's intentions. Surface vessels likewise were out. Submarines? It was at this stage that Courtney Whitney began to benefit from the earlier successes of AIB infiltrators and from the alertness of the encouraged guerrillas. When AIB had first proposed to insert agents via tactical submarines, our reception at Naval Headquarters had been a cold one; understandably so, but nevertheless cold. With such a mere pittance of submarine force, Task Force 77 did not relish taking out one torpedo for every man and gear we proposed to send north. It required a direct order from the Commander in Chief to make this precious space available. But one day the situation changed as if by magic. The trick was accomplished quite unintentionally by AIB; a routine report by one of Charlie Smith's agents telling of the torpedoing of an enemy ship near Davao did it. Navy was more than interested. Could AIB report these "kills" as a part of its normal function? Indeed, we replied, puzzled at Navy's enthusiasm. We soon understood. Compelled by the shortage of submarines to demand strictest "hit-and-run" tactics by its submarine commanders, the Navy was unable to credit these thoroughly deserving undersea heroes with sinkings they claimed because the submarines were not permitted to remain in a position to observe, and thus verify, a sinking. But AIB's observers and guerrillas with the new radios could do it for them. Thus, for the first time, the undersea men were getting their richly deserved marks. There was no more reluctance to transport AIB agents, and, as far as possible, to stow supplies for the guerrillas.

But now it was desired to send in many more agents and much more supply. It will be remembered that the submarine which called for Read's first batch of evacues New Year's Eve was a giant. She was the *Nautilus*. COMSOUPAC needed her yet, but possibly the Navy could assign her sister ship, the *Narwhal*, to Brisbane for AIB dispatch; she was just as big, just as eager, and just as tough. Later, probably the *Nautilus* as well.

Thus it ensued that when Chic made his next trip into the Islands, he went via the *Narwhal*. The big "sneaker" was heavy with guns, ammo, carbines, grenades, radios, medicines, money, food—a thousand morale-builders for the guerrillas, a thousand morale-destroyers for the enemy. Not just seven or eight tons this time—instead, fifty tons. Whitney's section levied heavily upon AIB's personnel and facilities to do this; the Bureau responded in a manner that made Chic and Whitney grin in anticipation.

For this particular "show," however, a severe strain had developed owing to a belated order to prepare a party for insertion in the very heart of the enemy stronghold. Major J. H. Phillips, a onetime Mindanao planter, was to establish something similar to a Northeast Area Coast Watcher-type unit at Cape Calavite on Mindoro Island, just south of Manila itself. He was to emulate the Villamor type of organization and avoid guerrilla entanglements until much more was known about them in that area. As always, time was pressing, but in this instance it was so insistent as to make training almost incidental. Perhaps we had been lulled by the successes Chic had experienced. Possibly we had been overcautious with "Planet"? I thought not. My protests were listened to and some training programs were initiated. But *Narwhal* was waiting...

Aboard the vessel Chic continued such training as could be carried on; the big, dark-complexioned, friendly Phillips fully cooperating. But all too soon the *Narwhal* nosed into her landing on Mindoro. Radio advances indicated that the area was clear and that a "reception party" would be waiting to help unload.

There was.

While such a hearty celebration might have succeeded in the far-south Mindanao where Fertig had reasonable control over men and events for limited periods, such a public event simply could not be justified under the very windows of the enemy headquarters, as it were. Phillips was doomed the moment he set foot ashore, although it would be some time before the *Kempei Tai* drew its noose tight. The story was told that among those watching the landing was an inconspicuous Mindoran who slipped away without speaking to anyone. In due time he appeared at the Manila *Kempei Tai* as a seller of fruit. He sold his basket without trouble. In that basket he figuratively delivered the head of the genial Phillips.

The enemy needed only confirmations of exact locations. These were forthcoming from Japanese radio triangulation experts who had zeroed in on new and powerful signals, signed "ISRM." To the listeners, these were merely call letters. To us in Brisbane they were the initial letters of a famous slogan: "I Shall Return, MacArthur." Phillips' station was the western anchor of a chain that was hoped to be established clear across the Islands south of Luzon. On the east Charlie Smith would go in again to set up "MACA," which was the cryptic signature the Commander in Chief utilized on his interoffice communications.

In Manila, the Japanese radio monitoring experts smiled their satisfaction. The new station was very prolific in its traffic; in fact, it was so busy transmitting that it seldom paused to change the key words for its cipher system. Convenient for the crypto-analysts, because any system, however "tight," would yield if there was enough substance to work on.

This circumstance was quite as evident in Brisbane as Manila, but AIB took no comfort in it. The shock was a double one because it was apparent that the big-hearted Phillips had become deeply affected by the deplorable plight of the guerrillas and had forgotten secrecy in his determination to better their situation.

Emergency sessions were called at Heindorf House. By now the astute, scholarly Dr. J. R. Hayden, one-time vice-governor of the Philippines and an authority on its political economy and its people, had joined GHQ and was assigned to Whitney's section. The three of us paced the famous "Back Room" and pondered possible solutions. We were triply concerned because concurrently with the Phillips' developments a most secret plan for the evacuation of certain highly influential, thoroughly loyal Filipinos in acute danger in Manila had been going forward. These people were to report one night to a certain point south of Manila. There would be a hidden small boat. At the proper signal the boatman would emerge. The civilians would embark and push off toward a tactical submarine diverted for the job and waiting on the surface—that is, if all was well.

It was not. One evening at this delicate juncture KAZ, Darwin, could not raise ISRM. The report of "no-contact" came to Brisbane, and there was a little ripple of apprehension. Still ISRM had gone off the air before when hostile movements were sensed in the area. But Lieutenant T. Crespo of "Peleven" party on Mindoro had reported alarmingly that ISRM's position was common knowledge.

Cautiously other net-control stations in the area began to call, not only to ISRM but to some of the satellite stations Phillips was known to have established along the coast. No replies. It was believed best for KAZ itself to refrain from calling so as not to betray our anxiety to the enemy. KAZ monitored instead, day and night. AIB's Dutch station hidden near Darwin changed coils to enable it to switch from the Netherlands area to the Philippines and listen. Nothing. The Navy's powerful receivers monitored.

The total "catch" was a terse hint from Peralta on Panay. He was trying to confirm a disturbing report that had come in by runner.

A long time later—a long time after we had learned the unhappy truth from our own sources—an official Japanese intelligence document, Report B, No. 51, covering the activities of Watari Group in Luzon during the first fifteen days of May 1944 fell into our hands. Paragraph 3 of the Summary contained a familiar name. The Allied Translator and Interpreter Unit operated by that veteran intelligence officer, the pirate-featured Colonel S. Mashbir, worked on it.

> 3. It is certain that the American secret agent, Major Phillips, was shot and killed. Captain Esugera, C.O., of Bataan and Cavite combat areas, who is now under arrest, was a subordinate of Major Phillips. When he met W.O. "Waisu"

[presumably Warrant Officer B. L. Weiss, Phillips' radio operator], an American agent engaged in radio work on Mindoro Island, the latter said he had heard confirmation of Major Phillips' death. Phillips was an Army agent who originally entered Attu by stealth—statement of Phillips' liaison man, a Filipino.

What this report didn't mention was that the *Kempei Tai* had struck suddenly in *three* places simultaneously: Phillips' headquarters; a Batangas Coast Watcher station; and a point midway between that place and a spot where the submarine was due to surface and take aboard the group of ten prominent civilians. Shortly afterward there had been another swift descent—on Elena Apartments in Manila where some of the civilians had lived. The torture dungeons of Fort Santiago were becoming well tenanted.

It was also much later that a certain Japanese soldier was captured in the Hollandia area who had seen duty with the Manila *Kempei Tai*. Yes, he had been on armed patrol under secret instructions in the Cape Calavite area about February 26. Yes, there had been an action. An American secret agent had been surprised while bathing. He had leaped for his arms, but had fallen at shots from the Japanese. There had been other sharp engagements with a considerable number of guerrillas. The American had escaped. He had thrown money away; in one place the *Kempei Tai* leader had found documents, cipher material, he thought it was. No, he didn't think Major Phillips had died then, because they found a blood trail, but it was lost eventually. He did think that the wound had proved fatal within thirty-six hours. Later he had heard persistently that such had been the case.

And what about the Japanese launch that suddenly appeared and captured the Filipino party in the small boat off Calavite a little after? Had there been any connection? The Japanese PW said no. There had been a list of names captured in Manila some weeks after Phillips' death. Well, then, he was asked: Had someone in Phillips' party talked?

Ah. There had been one man captured in the raid of the secret Coast Watcher station in Batangas. It was one of Phillips' men. He had, ah, found it expedient to talk. Yes, he had talked. But what he had said exactly the PW had no way of knowing. The Japanese believed in a policy of quick rotation of *Kempei Tai* patrol members so that none ever knew too much of one area and its operations.

This pretty much substantiated information brought down in person by Weiss. Miraculously he had escaped the ambush of February 26. It had been agreed, he said, that in case of surprise, headquarters staff members would flee in different directions to minimize the danger of a wholesale coup by the enemy. He made his way to Panay, reported to Peralta, and eventually was evacuated by submarine. (In September of 1944 I again said good-by to this courageous young man who, despite his harassing experiences, requested that he be returned to the Islands. Accordingly, he went in with the submarine *Seawolf*, a converted operational vessel now used for cargo work and assigned to Whitney's Philippine Regional Section. The *Seawolf* carried Weiss and fourteen others for Samar. She never arrived. There is some evidence that she was bombed out of existence by friendly aircraft which by then were raiding the Philippines.)

Various other members of the original Phillips party eventually were accounted for. Some reported to other guerrilla leaders, and one reported to the brilliant Commander G. F. Rowe of the Navy, who was sent in subsequently by Whitney to replace the ill-fated group at Cape Calavite. (Rowe, with splendid radio, photographic, and weather equipment and tested personnel, was to perform in an outstanding manner right up to the time of military operations in 1945; he refrained from involvement in guerrilla affairs as much as possible.)

Chic was deeply affected by the Phillips disaster. By this time he had done more, much more, than his share and he would have been justified in taking a base job. But the memory of the brave, possibly foolishly goodhearted Phillips drove him on without recess. This coin-flipping, merry-eyed, chuckler-at-fate went into the Islands on not fewer than half-a-dozen missions, each one of which broke into multiple missions once he was inserted by submarine. During one of these missions, in mid-1944, he made contact with Luzon guerrillas on the east coast, taking with him Private Courtney Whitney, Jr. Young Whitney was charged with delivery of certain messages for the guerrillas. Both he and Parsons returned on the same submarine that took them in.

Chic also is credited with making a momentous landing on Leyte in October 1944, just prior to the invasion. Through his intervention, at utter disregard by him of his own safety, the town of Tacloban was spared both bomb and navy big-gun shelling when he had found that to destroy it would have worked the greatest punishment upon the Filipinos and none upon the enemy.

That the messages he subsequently managed to send back to the Fleet and to Advance GHQ were the pivots upon which detailed invasion operations either were spun to conclusion or held, or altered, there is no doubt.

And yet here, on Leyte, the scene of his greatest triumph, Chic came closer to death by Japanese hands than he ever knowingly had done before. He told me about it in a shot-riddled building on the Manila water front after the Japanese had been pushed back onto Honshu itself.

"The chances were at least eighteen real ones to nothing flat that I was a goner," he said. "I was in my usual 'business suit' of old cotton pants, dirty and ragged, and a shirt to match, while on my head was an old saw-edge straw hat. I was barefooted. I'd sent two guerrillas on ahead as usual, while two native farmers were with me. The guerrillas stopped at a main-road intersection, and I concluded from their general behavior that the coast was clear. We went on—right straight into a patrol of Japanese soldiers who were carrying every kind of weapon except atomic bombs.

"Well, for all the action I was capable of at that moment, they could have sent a sick Boy Scout to take me. I was absolutely paralyzed with fear.

"And that," grinned Chic, spinning a well-worn Australian florin into the air and catching it expertly, "was my salvation."

Chic stared at the coin a moment, then went on.

"They slogged by in a ragged file—eighteen of them, so close to me that one or two actually had to lurch to the left to avoid bumping into us. We just stood there, frozen, my arm on the shoulder of one of the farmers, and stared hard the other way.

"Every second I waited for iron hands to grip me, or a bayonet to stick into my tight belly, but—say! What do you know about *that?* This coin came down tails, and it shoulda been *heads!* You know—I think I'm still scared—just *thinking* about it."

Sulu Sharpshooter

ONE CONSEQUENCE OF THE GALA EVENT THAT characterized the landing of the Phillips party on Mindoro was that the big *Narwhal* became the object of the enemy's concentrated attention. For a prolonged period following Calavite she seemingly dropped out of existence. AIB had no word, the Navy had "not a tinkle." The facetious phrase had nothing in common with the true state of feelings at Navy. Tension built up by the hour. Days and nights of it. Finally a message did come in—from the Japanese.

Intercepted, this fragment intended only for Japanese consumption, had included in clear language the word *Narwhal*.

Again there was concentrated work by Mashbir's translators: *Narwhal* had been observed by the enemy off Calavite, *proceeding southward*.

Then she was safe.

But it was obvious that her continued welfare was a matter of hour-to-hour concern; the enemy was determined to kill her. This had double importance at this time because Whitney and AIB were concerned in another deeply confidential deal to effect the rescue of the American men, women, and children of whom Villamor had heard when he debriefed Dr. Bell of the Silliman Institute. This little group, always one short hop ahead of a determined enemy, could not hope to elude him forever. There also would be some military personnel requiring evacuation. The rendezvous was set and *Narwhal* was to collect them—if they, the enemy, and the submarine could play their hide-and-seek game to a fine point just a shade in *our* favor.

Time after time quick alterations had to be radioed north to counteract equally quick moves by the hunters. And then it was done. Now *Narwhal* with her passengers stowed inside was on her way to Tawitawi to drop some sorely needed supplies to Captain Frank Young before turning south to Australia.

Would it be from the frying pan into the fire? Young had radioed that treachery was abroad in the Sulus, but that he had posted agents everywhere to try to determine if news of the submarine's expected call had leaked out. At GHQ it was felt that if treachery existed, Young's own shrewdness, his own steel-trap mind were the best counterfoils. Instructions went out to the *Narwhal* giving time and place in the Tawitawis. After that she would report to still another rendezvous off the northwest coast of Australia where her previous cargo of passengers would be transhipped to an Australian vessel which would take them to a secret landing spot.

Frank Young had been a third lieutenant in the Philippine Army that had rushed north from Manila to meet the overwhelming invasion of the Japanese pouring into Lingayan Gulf in December of 1941. Together with the American Thirty-first Infantry and other Fil-American elements, they had fought, been cut off, maneuvered, and fought again, but had always been compelled to retreat before the enemy hordes. The situation grew desperate. The main American defense forces were pouring into Bataan Peninsula for the last stand. But Young could not make it; he was wounded and in a field hospital that was being abandoned. He escaped. For months he ranged as a guerrilla under Colonel Claude Thorp. Then Colonel Thorp gave him some messages to deliver to General MacArthur. To reach Corregidor would have been a real feat in itself. But the Commander in Chief was no longer on the Rock—he was in Australia.

Young considered: he had his orders and he had his legs. But he would need a boat. He slipped down through the maze of the Islands until he reached the southern group. Here he joined with a keen-minded German who said that he hated the Nazis and their allies, and that he, too, had information he believed MacArthur would be glad to possess. They would go together. Young acquired a Moro-type craft with a single ragged sail— together with six silent, able Moro crewmen. They set off.

Young had long since discarded his uniform. Except for his height, he was as much a Moro as his crew: lank-haired, blackened by the sun, attired in ragged shorts and a dirty shirt and a conical native-type hat. The Moros did not like the German and, in their simple, direct way, occasionally tried to kill him. But Young was alert, day and night: he knew that the German's knowledge of navigation and his little compass were all that stood between them and extinction on those sun-beaten seas. They made their landfall late in 1942 and were rushed to Brisbane, Moros and all. Young was asked about the messages from the American colonel. "Oh, yes, sir—the messages. I have them here, sir." From his hip pocket he pulled well-creased, somewhat tattered papers.

Their arrival marked the opening of a new and salty period of learning for us in AIB. The presence of semi-savage Moros in "white Australia" offered its problems; fanatically devout Moslems in an all-Christian community provided others. To train them and Young in the fine points of the white man's idea of stealth called for exercises that were never, never boring—such as the night Young decided to test the proficiency of his charges.

He attired them in the jungle battle dress complete with helmets and carbines and told his pupils that they must deliberately let themselves be seen by such Brisbanites as might be abroad in one of the suburbs at that time of night. Then they must "melt" and must successfully elude the police for the remainder of the time required for them to traverse the entire city and report back to him. The only element lacking was that Young had neglected to inform AIB of some of the fine points concerning his "guys." The ensuing hours were made very lively for the police. Their telephones were jumping to calls by thoroughly panicky inhabitants convinced that the small, dark soldiers they had caught a momentary glimpse of were the first of the long-feared invasion. It is only right to add that the police, in turn, eventually made things lively for AIB.

"Tenwest" party embarked May 23, 1943. Headed by the grave, thorough Jordan Hamner, the party was to split up after it landed on Mindanao. Young's half would slip into the Tawitawi group and set up watcher units to report on possible southward movements by the enemy, especially through the key Sibutu Passage. Hamner was to continue westward and eventually try to effect junction with a British group of agents from Mott's section of AIB which had been working up the Borneo coast. Together they were to effect a solid observation of the sensitive Balabac Straits between Borneo and Palawan Island of the Philippine group northward. Thus it was felt that enemy movements either down through the Islands and into the Celebes Sea, or from Singapore through the Balabac and into the Celebes Sea, would be detected in time to offer warnings.

From the beginnings of modem times, however, the Sulu Archipelago has been a problem area for the white man—and for any non-Moslem. The Moro was, and is, uncompromisingly hostile to the "infidel" of *any* color, even his own. Little wonder then that Young had to report "treachery" abroad in the Tawitawis. Even so, the combined cunning of the hostile Moros and the maneuverings of the Japanese failed to silence the signals of Station "FGQ" hidden in the humid dark hill country of the Archipelago.

In fact, it was through this station which Hamner and Young had inserted with the help of the guerrilla leader Lieutenant Colonel Alejandro Suarez that the first intelligence information was relayed to Australia in mid-1943, as to the fate of the Australian Eighth Division, unheard of since the fall of Singapore early in the year. An intrepid band of Australians from that unit had managed their escape from a prison camp at Sandakan, North Borneo, and had reached Tawitawi Island where they had been nursed back to health by Suarez's men. (Two of these Australians, "Jock" McLaren and Rex Blow, later performed incredible feats while fighting with the American guerrillas on Mindanao, and as though that were not sufficient, in 1945 McLaren reported to Advance Headquarters of Allied Intelligence Bureau on Morotai for missions of utmost danger in the many operations conducted by the Services Reconnaissance Division unit of the Bureau against Borneo.)

Hamner had gone on, carefully, purposely. But this was one assignment that even the intrepid Parsons likely would have voted down; it is doubtful if Chic would have considered it feasible to send in a white man at all. I must assume my share of responsibility for that error. In the second place, experience was to teach the fallibility of trying to work a joint enterprise between two separate sections of the Bureau with their different methods of operating, their unrelated systems of cryptography, and the contrasting temperaments and motivations of the personnel.

The two parties reported that they were unable to make their rendezvous. In addition, Hamner's health had not been recovered to the extent it seemed to have following his prolonged exposure with Smith in their epic small-boat voyage from the Philippines to Australia. His teeth became infected. Then a fungus developed in one ear. It was plain that Hamner would have to be evacuated and another party developed for the Balabac watch. To take the heat off the guerrillas in the Tawitawi area who had been doing their best to protect him against the Japanese and Moros alike, Hamner struck westward and established a secret command post on Borneo.

This move made him inaccessible for evacuation, for as has been seen *Narwhal* was making deliveries in the Tawitawi area and had the full nature of the grave situation confronting Hamner been realized sufficiently early in Brisbane, he would have been directed to effect junction with the big submarine when she came in to unload. Instead, he was far away. Despite his wretched health, he would stay with a rough situation doing what he could and relaying information until he was evacuated in March of 1944.

For us in the "back room" at Heindorf House in Brisbane it became the old familiar "sweating out" period between the last radio contact and the "rendezvous effected" message that should follow once the submarine had gotten safely out of the area and could break silence again.

But this time the message did not come. The silence finally was ended by a flurry of maddeningly corrupt ciphers from Young. To make it worse, his transmissions were so weak that frequent repeats were necessary. Obviously he was not sending from his main-base set, but from some portable unit in a location that might be good for his security but bad for radio emissions. It spelled trouble all around.

Days later I was off in the Timor Sea to get the story.

Treachery and betrayal had kept twin watches in the Sulus. All had gone well with the contact initially. Lieutenant Young's fleet of small boats had swarmed up to the big submarine, whose deck was piled with cases of money and equipment. Guns were manned, lookouts doubled. From the conning tower the penetrating eyes of Commander Frank Latta, United States Navy, had missed no detail. And Young's liaison transmitter and receiver on shore should catch any possible warning of danger from the main station in the hills.

None came. Yet, suddenly, Latta saw two destroyers bearing down on him from opposite ends of a nearby island.

"Crash dive!"

His shout was followed instantaneously by the blare of the klaxons.

The big sub shivered as her propellers engaged to engines that had been idling and were now at the "full."

Men were plunged into the sea. Bobbing off into the churning waters went scores of burlapped cases. There were shouts and cries and a panicky rush of small craft for land to escape the two ships racing down on the *Narwhal*.

A moment later only a whirling boil of foam and bobbing cases marked the spot where the big submarine had been. Then the cases were hurled violently aside by the slicing bows of the destroyers. From the fantails of the ships the depth charges shot into the air and down into the water. The sea heaved and the air shook with the explosions.

The destroyers canted sharply, reversed courses, and repeated their operation.

Slipping into the jungle, Young raced for his station in the hills.

In the *Narwhal* men, women, children, crew snatched anything solid. Some missed and went catapulting forward into the bulkheads, so steep was the dive. The submarine shivered with every detonation of the charges.

The *Narwhal* touched bottom. It would be deep enough—or it would be their grave. Latta tried not to think of the terrific pressure on the plates.

There had come an interval of respite. Then the detonations began again, each one more violent than the one before it. The *Narwhal* shuddered, lifted heavily, and settled at a sharp angle. They were left in utter blackness as all lights failed.

The *Narwhal* was not destroyed. Some days later I came aboard and listened spellbound to Latta's account. At the point in his narration when he told how the lights had failed, he paused for a moment, then went on.

"They stopped after that," he said quietly, as though he had been speaking to me of fish biting, or church bells ringing.

I stared at the calm waters of the Timor Sea sloshing against the rusted sides of the *Narwhal* and waited.

"You'd never believe what saved us," he continued.

With a sigh that was half thankfulness, half pride in the old *Narwhal* for taking the terrific punishment, he explained that the submarine had not buckled anywhere from the last series of explosions. Following these there had been a long cessation of depth bombing. Latta had not been fooled. He knew the enemy was listening.

"Pass the word," he had whispered. "Not a sound. Not a sound of *any* kind, understand?"

The enemy was using "dotters." This was a device for sending sound waves to the bottom of the sea and registering their echoed return. By studying the characteristics of the returned sound waves, the destroyer captains could roughly determine the outline of whatever structure on the bottom was reflecting the waves.

"It was an appalling strain," Latta related in that same un-dramatic way. "And then suddenly we heard them at it again. I could not understand. It was a tremendous bombing but it was not for us—or, rather, it would not *get* us." He chuckled. "The enemy had registered upon a submarine shelf of rock about three hundred yards away from us. The shelf was long and narrow. He fairly blasted it out of the sea. Even so, after he'd gone, we had some trouble making the surface because of the mauling we'd taken initially coupled with the distant but even more concentrated bombing."

The real nature of the betrayal never was known to AIB. Only a short time before, Young had escaped two ambushes and the following week he had put the blade to the neck of one whom he had trusted—and found trafficking with Moro agents of the enemy.

The attack was the *Narwhal*'s closest brush with death.

Young's position in the Sulus had rapidly become desperate. Denied his badly needed funds, armament, and drugs, he found himself forced to abandon secrecy status and openly called upon volunteers for a guerrilla force, declaring himself to be the leader officially appointed by MacArthur.

At first he was unable to gather more than a baker's dozen; in fact, enemy agents were having considerably more success at recruiting. The Moros were deserting wholesale to the Japanese.

Driven by patriotism and by his fanatical hatred for an enemy that had killed his own parents, Young faced would-be deserters with his most effective weapon—his flaying tongue.

General MacArthur's chosen leader, was it? thought the Moros. And only a lieutenant?

"I am *not* a lieutenant. I am a captain!" Young had shouted. "Appointed to lead you. A captain, yes."

A few days later there was transmitted to the Commander in Chief by Whitney one of the most unusual messages of the war. Paraphrased, it went like this:

I HAVE STOPPED A REBELLION SINGLE-HANDED. BUT I HAD TO BE A CAPTAIN TO LEAD THEM. DO NOT MAKE A LIAR OUT OF ME. MAKE ME A CAPTAIN INSTEAD. AT ONCE, PLEASE.

Whatever may have been the general's broader feelings toward this characteristically unmilitary message from Young, there is no question as to his reply, which he ordered sent at once.

Young was to be a captain, not only because Young said so, but because

General MacArthur said so.

It was unanimous.

And so it was that Captain (still later, on Korea, a lieutenant colonel) Young still was holding some sort of sway in a country that was death itself by the latter part of April 1944. April 20 it was, to be exact, when the first of a series of highly charged messages began to come through in a bewildering scramble of bilingual cryptography that would have defied the Black Chamber itself— had they not known Young as we did.

Providentially, we were able to get something out of it. The information was rushed up to G2, to the Commander in Chief.

Ten, twenty. . . twenty-seven ships now were in the Sulu Seas, reported Young. And more on the way.

His signaling became more erratic than ever. Brief messages that we hoped he would take the time to decode were sent to urge him to observe tightest signal security for now he was GHQ's only "eye" to see which way that formidable fleet turned, once it got to where it could go north, east, or south. Apparently he did decode one of them. To our vast relief he "Rogered" and held his tongue while he watched.

Then he broke radio silence in a manner that was Young to the tips of his long brown fingers. He sent his answer in clear English —Young style:

DOS GUYS ARE MOVING EAST.

What the American fleet eventually did to "dos guys" in the Battle of Leyte Gulf is a matter of history.

Philippine Snatch

ARRANGEMENTS TO RECEIVE THE *Narwhal*'s refugees into Australia and to rehabilitate them enough to enable them to proceed to their various destinations were practically as secret as those which effected their removal from the Philippines. There were more where they had come from. If concentrated reprisal by the enemy was to be avoided, it was most important that the Japanese be denied any hint that an evacuation program was under way, or that any part of it had been successful.

At Whitney's request the assignment fell to me to coordinate the operation involving the rendezvous with the *Narwhal*, the subsequent smuggling of the party through the vast military camp of the Darwin area without the knowledge of any save those concerned in the job, and finally the movement to Brisbane. From there the Philippine section had arranged for their rehabilitation at a coastal rest area north of Brisbane, where their presence would cause no particular comment.

The first stage involved our flying north to Darwin in two C-47's quaintly called *Cold Turkey* and *Hairless Joe*. There were two others besides myself. One was Captain Willa Hook of the Nurse Corps, whose courage, devotion, and dependability had been thoroughly tested in the fires of Bataan and Corregidor (she was one of the small party of nurses who had been air lifted to Australia at the last moment). Since a doctor might be needed as well, Major H. Eldon of the Medical Corps was the other.

Eldon and Hook had been surprised upon boarding the aircraft to note that in addition to the necessities and "comforts" accumulated for the evacues there were several tons of solidly-packed gear with no markings.

But they had been warned to see nothing and say nothing except as concerned their own mission. The stuff actually was the final installment of radios and other gear for a new AIB party destined first for Mindanao, and then northward to effect that eastern anchor of the string of secret stations planned for the whole width of the Islands "north of twelve degrees." Phillips was to have been the western end; Charlie Smith would establish the eastern anchor in the very sensitive area of the San Bernardino Straits, one of the two principal ways of entering the Islands from the Pacific side (the other was the Surigao Straits in the south, which Parsons already had "planted"). The party itself already had been flown to Darwin and was installed in isolated billets at what AIB termed "Lugger Maintenance Station," on the coast northwest of Darwin proper. It would be at this same secret installation that we proposed to land the evacues from the sea side, install them in rest billets for the night away from the Philippines-bound party, and then smuggle them out at dawn the next morning. After that, the Smith party in turn would load and board the waiting *Narwhal* and disappear northward. Neat arrangements—if everything worked smoothly and on the dot. There was one nagging concern: someone "in the know" had talked: Nurse Hook had received a phone call in the middle of the night before we had come for her to go to Archerfield to board *Cold Turkey*. The caller had been a newspaperman who said that he understood that she had been purchasing rationed clothing for a "two-year-old who was an escapee from the Japanese, etc., etc." She had cut him off and run for the car.

After a daylong flight during which the aircraft seemed to stand almost motionless above a rust-colored land where time itself hung suspended, we put down at Darwin and were driven in Australian light trucks through the scattered buildings of Darwin, many of them unrepaired after the heavy bombing of early 1942, and out into the bush. There the trucks stopped. Ahead were the bunched shadows of foliage topped by the straight boles and thin cover of the ever-present Australian "gum" trees. Our escort and commander of the station, Captain Jack Chipper of the Australian forces, jumped out, his boots raising red dust as he landed. Via a telephone protected by an "elephant iron" chalet, he notified the guard rings of our approach. We moved slowly over a trail flanked by foliage to a gate. In the gloom we could see the multiple wires of the fence "in depth." Here and there against the pale light were discernible sinister black objects like spiders in the maze of the fence—booby traps.

Late the next afternoon, after we had finished arrangements for receiving the evacues that night, we returned to Darwin and boarded an Australian Navy special-purpose vessel.

The skipper, Lieutenant Commander Erricson of the Australian Navy, waited only to check us in, then gave the order to cast off. The word from the American naval senior officer in the area, Commodore Jack Haines, had been to the effect that the *Narwhal* would fulfill her "expected time of arrival" at the rendezvous.

Night began to fall. The *Chinampa* pushed on steadily. Erricson continuously studied the horizon through binoculars; it was important that the meeting be effected while there still was enough light. If calculations had been right, there would be.

Then we saw it—a mere black dot on the now-indistinct blur of the sea. The *Chinampa* payed off to starboard. The *Narwhal* drew near. Now her conning tower was plain, her squatting guns—and even tiny figures clustered on her deck. She lost way and waited for Erricson to close up in the most favorable position. This he did with a fine show of seamanship. Against the somber gray of the submarine bright spots of color marked the presence of the women, incongruous in this scene of battle-tested efficiency; for the *Narwhal* did not forget for a moment that this was war and that she was designed to fit a pattern of destruction—all her defenses were manned, and at her slender stern crewmen hunched over a hydrophone station.

Now the refugees were leaving the area amidship and going forward in anticipation of the transfer to the vessel that would bring them to land— safe land. In their faces were plainly to be noted the indelible etchings of nearly two years of constant apprehension and flight, yet outwardly, at least, there was only moderate indication of physical disability. It seemed there would be little work for the medicos.

The ships were lashed to each other, the word for transfer given.

Without hesitation, and flinging their good-bys and gratitudes over their shoulders, the refugees stepped across the gangplank joining the two vessels, and clambered over the *Chinampa's* rail—except one.

Her face was pinched with suffering. Her thin arms bespoke severe malnutrition. She was seated in a chair on the submarine's deck. Beside her was a dark-eyed, quiet youth.

"We'll just carry her aboard, chair and all," called one of the sub's bearded crewmen. "Here we go."

In a moment Mrs. Evelyn Birchfield was snug aboard the *Chinampa*.

Beside her stood James Birchfield, whose fifteen years of strength, although taxed, had served him to carry his mother about in his arms, much as she had carried him years before. All through the voyage she had sat in her chair. So solicitous had been every member of the crew from Commander Latta to the newest able seaman that the seven days and nights had fled by leaving her with many memories of human kindness.

The vessels were unlashed, the *Chinampa's* engine-room telegraph jingled. The dark water between the two ships bubbled and swirled from our propeller.

As we saluted the *Narwhal*, the air was filled with farewells and expressions of thanks, hopes for reunion, and prayers for luck. Of the names called out by the submarine's crew, one was predominant.

"Good-by, Stevie." "Good voyage there, Stevie." "Stevie, don't forget me, and so long."

Who could this be? Someone explained to me.

"Little Steven Cryster, there. Believe me, he was the real boss aboard the submarine. It was his ship."

In the quickly-gathering dusk his blond curls and big blue eyes could just be discerned. Truly a beautiful child of two years. His stout little arm pumped his farewells, but he was too overcome to be articulate at that moment. He was in the arms of a somber-eyed, dark-haired woman, Mrs. Glenda Hallea Cryster, his mother.

Then, as though by common agreement, there was silence as each withdrew into his own solitude at this moment of departure. They had seen that vessel rise out of the sea to come to them. After being pursued for two years by misery and hatred, they had hardly dared to hope. But she had come. Life was to be lived again. Ahead was a new land, a hopeful land, where there would be friends, American soldiers—thousands of them, fresh from America; airplanes with the United States Air Corps insignia.

I thought: here at last was tangible, human evidence of the result of GHQ's constant efforts. Almost one year before for us in AIB these efforts had been launched with the departure of the first party under the intrepid Villamor. One's mind blurred to the montage of many scenes, many words, anxious days, endless nights, halting messages, mounting hope, constant apprehensions because of the seemingly insurmountable odds.

Now in the *Chinampa's* engine room the telegraph gonged for full speed. We had swept in a huge circle and now were headed west-southwest. The lights that marked Port of Darwin would appear soon.

"What's the course?" Erricson asked Captain Chipper.

"About two more points to starboard. A single light will appear within five minutes."

The skipper studied the shore with a night glass. The headland we had rounded lay flat and dark, like a giant crocodile's head and snout, motionless upon the sea. The possibility of some hitch in the arrangements intruded into the mind. The skipper continued to study the shore.

Then in exactly the position Chipper had predicted a yellow eye gleamed briefly in the crocodile's head.

"In just a moment another light will mark the limit on the port bow," explained Chipper.

Punctuating his sentence, a second point flickered, then was steady.

"Pay off two points to starboard," suggested Chipper.

"Two points it is. How far do we go in?"

"Until you pick up a trawler off the starboard quarter. She'll flash once. We answer the same way. Then reduce to half-speed."

For several minutes the only sounds were the steady exhaust of the engine and the lisping of water from the bow. Then a searchlight beamed across the sea and held us fast for a brief moment. Our engines eased. And our own searchlight picked up a small trawler, or cabin cruiser. She was standing by. The *Chinampa* lost way, and nudged close to her.

We moved onto the cruiser, which brought us to shallow water. There we were met by dinghies and rowed to shore.

Now in curious procession we made our way by the light of oil lanterns placed along the trail winding up to the main station house. On the veranda was a row of nine bunks, each whitely collared in snowy, turned-back linen. And in the middle of each a Red Cross bundle.

There these weary people stopped. Tired eyes ranged down the line of waiting comforts. Slowly they went forward, each turning into a space between cots. Then a soft, hesitating voice . . .

"It's . . . it's just like Christmas . . ."

The tension dissolved. Tears were unashamed.

"Look. . . Look. . ." said one woman. In her hand she held a nylon tooth brush, sterile and new in its cellophane case. Her voice trailed off. Her lip trembled. She turned, laughing, to Nurse Hook. "I know it's silly, but... The back of her hand pressed hard against her lips to stifle the threatened hysteria. Nurse Hook's eyes held hers and she winked. Now the laughter was normal, healthy.

They were all there—quiet Donald C. McKay, whose mining genius had developed the Mindanao mother lode, and attractive Mrs. McKay. Beside them, soft-mannered Mary, their daughter. Grizzled Captain John Martin and Mrs. Martin. Mrs. Cryster, Mrs. Helen Welbon, who together with Mrs. Nellie Varney had come out without their husbands, sorely needed by guerrilla commander Colonel Fertig. And Mrs. Stanley Briggs, whose husband would not come out, then or ever. There was Mr. G. E. C. Mears, a British subject (who had married a member of the United States Army Nurse Corps, Lieutenant Robertson). In a nearby room was Mrs. Birchfield, and beside her the ever-faithful James. There were two other civilians, but they were bunked with the servicemen in nearby huts—and two loyal Norwegian sailors who had been rescued from a torpedoed ship.

After a pre-dawn breakfast there was another roll call. In addition to the civilians, there were Major Halbert D. Woedrugg, Sergeant Frank Duff; Master Sergeant Albert Kirby, Jr., Corporal Cyril Grohs, and Privates Rommy Stewart, Oscar Smith, Frank Harayda, Leo O'Connor, and Aldo Maccagli of the Army (mostly Air Corps), Lieutenant Colonel Justin C. Shofner, Major Jack Hawkins, Major M. Dobervich, Corporal Reid C. Chamberlain (whose story was a chapter in itself) of the Marine Corps, and George W. Winger, Otis E. Noel, and John L. Houlihan, Jr., of the Navy.

We climbed into military ambulances. That was one way to smuggle women and children down to the airfield without observation, even by inquisitive sentries. The sun had not yet risen as we caravaned past the gutted buildings of Darwin airdrome and onto the strip. We descended. Faintly discernible in a nearby revetment was the spreading shape of an airplane.

"A Liberator!" exclaimed one of the evacuated Air Corps personnel, joy in his voice. "Just let me *touch* her!"

Men, women, and even tiny Stevie followed his lead. An American airplane.

At eight o'clock on the dot we were airborne.

High aloft in *Cold Turkey* and *Hairless Joe* the inevitable reaction set in and most of the party slept. At one point during the long journey to Brisbane I peered out of the flight-deck gangway of *Turkey* to make a check. A man was just settling down close to his wife whose face in this slumber relaxed into grim lines of wear and fatigue. With infinite gentleness he placed his arm under her head to afford her a pillow.

These would be the first of some four hundred and fifty to be "snatched" from the Philippines in this manner before the end.

Mindanao Mender

IN THE DARK HOURS BEFORE THOSE AMBULANCES had lurched off to the airfield I had slipped away from the main lugger station building to pay a farewell visit elsewhere on the station. In the steamy darkness of another dormitory Charlie Smith had shaken himself into wakefulness and with him Captain (later Major) James L. Evans of the Army Medical Corps. There was a last-minute broad check over of their plans.

They would travel together to Fertig's headquarters. There Evans would, first, endeavor to weld Fertig's radio net into an efficient, dependable unit that would function at all times despite the combined attritional factors of inexperienced personnel, the jungle, and the enemy; second, he would accord such limited medical help as was possible to guerrillas and civilians, so long deprived of such help. As for Smith, when conditions were reasonably favorable he would work his way northward to establish that eastern anchor for the trans-Philippines chain of agent stations. His story will be told later. Meanwhile, Evans...

An odd combination, Evans. A mind that sometimes seemed to exude the essence of pure intelligence. One felt it. An orderly, analytical mind which properly questioned all that was brought before it, and having queried sufficiently for proper identification, filed the matter away for eventual use. Holder of awards in English composition, Evans was an equally accomplished youthful physician and surgeon. An essential friendliness, a gift for gently prodding into the intricacies of the human mind and heart marked him out as a natural psychologist. Yet no more of rhetorician and doctor was he than he was communications expert.

Literally a triple-threat man for AIB.

Wordlessly we gripped hands. What would happen to him before we might once more shake hands—or would we, ever?

What did happen to Evans combined to make one of the most fascinating tales to emerge from a record crowded with the dangerous, the daring, the bizarre. Yet there were lighter moments. One of these occurred not long after the *Narwhal* had taken him aboard and begun the run back into the Islands. Half-blinded by looking toward the early-morning sun, Evans had so quickly obeyed a command to "clear the bridge" that he missed every rung of the steel ladder and landed below with the speed and force of an unimpeded falling body. "Evans' Leap" became legendary in the wardrooms of submarines in Task Force 71 as the fastest known way to "clear the bridge."

Thanksgiving of 1943 was celebrated aboard the *Narwhal*, and was followed by the dangerous passage of the Surigao Straits. The *Narwhal* ran through at night, just awash, and submerged at dawn. She was at the end of her northing and the course was now southwest.

Unable to sleep during the ticklish run through the Surigao, Evans was up at dawn the next morning and joined Latta at the periscope. The commander straightened up, indicating that Evans should have a look.

His heart leaped at the beauty of the scene revealed to him. To eyes accustomed to white steel bulkheads, glaring electrics, and the colorless fixtures of his subsurface prison, the lush, unreal green that startles every beholder of the Philippines for the first time seemed to him something he was witnessing in a fantastic cinema. Vividly chromatic, extravagantly green, with the morning's sun rouging the highlands behind.

The *Narwhal* settled to the bottom to await evening. Then she went to periscope depth again. Latta invited Smith and Evans to have another look. They saw a white target on the beach. Then from the darker band of the shore above the target came three winking lights, doused and repeated, to be followed by darkness once more. It was the "all clear."

The *Narwhal* pumped the sea from her tanks and stood solidly on the surface. A barge came alongside. In it was a guerrilla officer named Money. He had with him a 3BZ radio transceiver, one of the units that had been sent in to Fertig previously. This portable set was tuned into the master control station inland and that one in turn was monitoring all of the little ATR4A's hidden along the north coast which would send in their alarms at the first sign of threat approaching the lair of the *Narwhal* and her precious cargo of radio equipment, guns, and medicines. The hatches were opened. They had arrived.

A few minutes later the bearded Colonel Fertig himself came aboard. Evans studied this lean, quiet man who wore that undefinable air of a commander and wondered if his country ever would truly realize what a debt it owed him—or would ever honor him in proportion to it.

Evans, Smith, and their radio operator, Robert Stahl, soon found themselves with Fertig in a launch that moved along the coast and into the Agusan River. Just before midnight the little craft stopped at a barrio called Ampara. The weary men pitched their jungle hammocks and slept.

After an early breakfast they pushed farther upstream in a smaller launch. There must be no time lost. Word would get around fast enough that the *Narwhal* had come in and discharged much mysterious-looking cargo. The job was to move the valuable radio gear far into the interior. Enemy shore-raiding parties who might turn inland would have to pass observers who then would alert the hinterland to danger.

There was nothing particularly secret about the headquarters of the most formidable guerrilla force in the Philippines. It was located in the small barrio of Esperanza at the junction of the Agusan and Wawa rivers. Except for local sentry rings, there seemed to be little in the way of warning apparatus. Nevertheless, it would be some time before the enemy in the coastal areas would know of it with sufficient accuracy to dispatch effective bombers and strafers to the area.

Evans initiated his communications survey. In a small hut at Esperanza he found another Teleradio which relayed to a powerful fifty-watt transmitter Parsons had brought in earlier. This was located in a well-guarded secret place in the hills westward. But relays were time-consuming and demanded air time for the enemy to monitor and triangulate. Surgeon Evans selected tools of a different type, probed the vitals of an ailing HT-9 transmitter and put it on the air with a special parabolic-type antenna that shot a concentrated beam at AIB's station near Darwin. The resulting signal was clear and strong. For the first time Fertig had direct communication from his Esperanza headquarters to Australia— that is, at nighttime. But Evans had other ideas. He wanted twenty-four-hour service and he wanted it without Japanese monitoring. He advised KAZ he was doubling his normal frequency and for them to prepare receivers that would take it. The frequency he was suggesting seemed fantastically high. But Brisbane's experts met the challenge. The result was nearly what Evans sought—and with no more than fifty watts for that tremendous distance. Normal procedures with no restraints on the kind of equipment to be used and no fear of enemy reprisals would have called for ten times the power to do the same thing on a twenty-four-hour basis. Evans was like that.

The accomplishment was only one short leap ahead of trouble. In fact, there had been only one warning of the swift enemy action that resulted in the loss of the Parsons' fifty-watter. This unit had been primarily to serve the Navy at Perth with immediate submarine attack data, and to relay Fertig's GHQ traffic. The cleverly-concealed position was uncovered by the sudden raid of enemy soldiers and the precious equipment destroyed. As a precautionary measure, the Esperanza unit was also immediately dismantled and hidden. Thus, no sooner had a peak in communications been effected when KAZ had to report a complete blackout.

Complete? Not quite.

Thin and uncertain, a voice had come through the air. Its call sign was "UU2." The message was relayed to Brisbane for study and identification. If Evans had accomplished a master stroke, one of those tiny, incredible Australian ATR4A's had done even more. To our astonishment the call was identified as being that of the station Charlie Smith had established near Davao in 1942 when he had gone in originally with "Fifty" party. Those two-and-a-half watts of power had spanned the whole distance! It was, of course, a freak and could not be expected to repeat. It did, however, every once in a while, as Charlie's observer sent out ship sightings originally intended for relay by Fertig.

In time, Evans came back on the air. The HT-9 had blown up for good in the wet tropical heat. He had substituted an Australian TW-12 which withstood the rigorous conditions. We had not yet learned to properly "jungle-proof" American equipment.

By early 1944 the traffic from the Philippines in general was beginning to assume impressive proportions. The Heindorf cryptographic section under Lieutenant C. B. Ferguson was a model of efficiency. If the messages were to be decoded, studied, and forwarded to the Commander in Chief with comments and recommendations in time for his morning sessions, it meant that Whitney had to quit his bed each morning at about three o'clock. Folk like Dr. Hayden and myself normally reported at eight o'clock to debate such issues as Whitney desired to try out on us. Then the paper work went forward to GHQ. By afternoon the replies generally were back for encoding. Sometimes the Commander in Chief acted on the recommendations as suggested, sometimes he amended them, and sometimes he quite ignored them. But the collective "batting average" of Philippine Regional Section remained gratifyingly high.

On the Mindanaoan end of the link was another cryptographic section, lacking no doubt some of the fine equipment of Brisbane's but, under one Harold Martin, nonetheless efficient. He was eventually commissioned in the Signal Corps.

Evans' net was "clicking." A typical Martin encoding from some of the stations hidden on the coast might refer to: "… medium cargo ship at 122.5 east, 12.06 north" or to a "convoy of small ships headed south of Sindangan" or "twenty-two bombers arrived Davao strip from north, are refueling." Maybe it would be "four three-inch anti-aircraft guns now installed in clearing at southeast angle of road intersection at [coordinates] with fuel dump hidden in trees five hundred yards north."

Ship information immediately went to Perth and Perth made its own contacts with submarine commanders. Obviously a direct tie-in with them would eliminate costly delays. Evans pushed for it. Previously Navy had declined to let the watchers have any access to the submarine frequencies. Now the proposal was urged with renewed energy, especially by Parsons. The Navy agreed. "Kill" counts immediately mounted.

One day a message was dispatched to Brisbane by Fertig in his own cipher. He had felt constrained to mention Evans' deteriorating health. The doctor had worked incessantly and despite urgings seemed unable to relax. Soon Fertig had the "pursuader" he wanted in the form of a message from Brisbane that Evans was to ease off on the now smoothly-working communications net and at his discretion make a medical survey in the area. Evans accepted the order amiably enough, but to suggest a slowdown for him was one thing, to effect it was another—regardless of the nature of his occupation. With the same devotion he had given radio, he turned to medicine.

With the help of a Filipina nurse, Evans set up shop in an abandoned house in Esperanza. It was a three-room establishment—a living room for him, a waiting room, and a "surgery." He added to his staff a one-time pharmacist's mate in the Navy—Henry Rooke. The trio was an immediate success. AIB came to dub him "the Mindanao Mender."

Business came from every direction—and in all forms and degrees of pathology and trauma. Digging Japanese bullets out of Filipino anatomy was interspersed with treating a constant parade of tropical ulcers.

One aged Filipina came from afar—for the word had traveled afar that a white doctor had appeared from no place at Esperanza. She had a large purulent abscess under one eye. Immediate excision was indicated. Evans signaled his nurse to prepare the scalpels. But the patient would have none of the knife. The doctor was troubled; the stinking infection would soon spread to nearby brain paths. He had an inspiration. A less-provident man, or one less susceptible to the urge of details in planning, might not have thought to include some of the new anesthesia, sodium pentothal, in his kits so hastily assembled in Brisbane. Evans had.

With much ceremony he directed that the incising instruments be returned to the sterilizer. Before her one good eye and the other affected one he displayed an intriguing little glass ampule. "Dream medicine" it was. Persuasive was his voice, flattering his words. Her suspicions dissolved. The "dream medicine" was pumped into her brachial vein. Soon sleepiness and relaxation in turn gave way to a slumber that left her insensitive to the swift work of the knife and the cleaning action of the irrigation solutions.

Ten minutes later the patient awakened, quite apologetic. It seems, she explained, that she had waited too long, and had fallen asleep. Then the sight of her operated face, protected with clean dressings, was framed for her in a mirror.

And from that moment the barrio of Esperanza and all the territory around belonged to Evans and his "staff" of two. The account of his prowess grew amazingly with every repetition. For Evans there was the added satisfaction accruing from the knowledge that, as far as anyone knew, it was the first time that sodium pentothal had been used in the Philippines.

There was another medical "first" during those days and nights when, unknown to them, the enemy was inching his way closer to a "find" of those stations which so brazenly were supplying GHQ with thousands of cipher groups every week.

In November of 1943 the medical supply depot at Brisbane had boasted a total stock of fifty ampules of the newly-discovered germ killer, penicillin. It was worth a small fortune because at that time the mass-production methods were still developments of the future. Yet the 155th Medical Depot had given up half of its stock to AIB.

Evans' patient was the wife of a civil official whose good will was important to the guerrillas. But it would have made no difference to him who she was, for in any case she was a human being whose life soon would terminate unless the acute infection in her cervical region was halted. The precious ampules were brought out, and in due time the house of Pajarillo greatly rejoiced. The age of miracles had not passed. To Evans came the satisfaction of knowing that, in addition to having brought happiness and health into a land where disease, poverty, and war's brutality had been rampant, he had to the best of his knowledge and belief made successful use of penicillin in the Philippines for the first time in medical annals.

Late in January uneasy whisperings began to come by way of the jungle telegraph. The enemy was "wise." It was nothing Fertig could lay a finger on, but real, nevertheless. Perhaps the enemy rdf's had at last pinpointed him? Or had there been spies?

He decided to act on his hunches. He instructed Evans to "fold" and go up the river to Talagogan; he would follow later and relocate his headquarters there; part of the equipment would be retained near the barrio of Pianing under Mr. Sam Wilson (whose "Wilson Building" had been a prominent landmark on the Escolta skyline in Manila before the destruction).

In the meantime another successful supply drop of bountiful proportions had been made by submarine. There was much equipment, new personnel, medicine—and even 20-mm guns. Lieutenant Monty Wheeler of the United States Navy had brought in a new naval transmitter to replace Parsons' smashed fifty-watter; he would operate the new unit for Navy.

Evans made the move to Talagogan and immediately set himself up in another "dispensary," together with Lieutenant Carlos S. Turla. He was too busy to be disturbed seriously by the latest rumors of enemy activity but got a good laugh from the report of an encounter of McClish's 110th Guerrilla "Division," which had just received one of the 20-mm cannons. The enemy had come in strong, quite unprepared for the ravages of the well-mounted field piece. At first they had been routed. But overwhelming numbers in time had the usual effect—the gun position was overrun. The enraged, unpredictable Japanese paused, then lighted a fire under the cannon, and retreated. McClish's men swarmed out of the jungle, put the fire out, reloaded the hot gun, and began a rapid firing against the soldiers' posteriors—with devastating effect.

It was a local victory.

But the over-all enemy command had experienced its fill of brazen submarine landings and now moved to mop up the whole northeast and east coasts. There would be no more landings in that area for a long, long time. Enemy attack planes swooped low over Talagogan, strafing and bombing. The little barrio became a shambles in which one man moved as if protected by magic—Fertig. He was not hit. But everything except one 3BZ Evans had buried was hit and reduced to junk. To our great relief in Brisbane, the sweet notes of this transmitter came through to give reassurance that they lived to fight another day.

Evans embarked on a program of hiding other 3BZ's. So perfect was the camouflage of these hidden stations that even the local operators had to memorize certain landmarks to find their way in to them.

But a new specter came to haunt them—hunger. The enemy's activities had all but severed the main supply line. Evans' already lean frame seemed to shrink visibly. The menu of *tankong* (fern greens) became so inevitable that the mere sight of the stuff set him to retching. Polyvitamin pills brought from Brisbane alone prevented the ravages of severe malnutrition.

How long could they hold out? It was all the more ironic now because the information coming from the important Davao area was becoming unbelievable in quantity and accuracy. The Illocano "natives" they had trained and sent in to observe that sensitive area had actually hired out to the Japanese, just as "Carabao Boy" had done in Manila. Mistaken for the ignorant farmers they represented themselves to be, these keen-minded youngsters, many of them college men with bilingual abilities, hired out for work in enemy ammunition dumps, airfields, and even in headquarters itself. Their gleanings were encoded by men of Fertig's command and others brought in aboard the last submarines to call, and transmitted by ATR4A's to the net control unit, then sent to Brisbane.

One message which described the presence at Davao of enemy naval units unsuspected of being anywhere near that area seemed sufficiently incredible even to the observer himself that he considered it advisable to append this line:

... I AM SOBER COMMA HALL.

Comedy relief, however, was almost as scarce as food in the hard-pressed Fertig area. And Evans, who burned nervous energy at an exceptional rate, had little more to burn. Fertig had seen the early signs and now he saw them in an aggravated form. If he was to preserve the man for future usefulness, he had to act again. The situation was too grave to permit proper rest. A change of scene, then.

Tired, thin, his nerves on edge, Evans pushed off with Filipinos in a baroto. Stowed in it was an Australian Kingsley receiver. Then they left the craft for a long trek inland. It was mid-1944 when they arrived at

Wilson's headquarters near the barrio of Pianing in northeast Mindanao. Gratefully Evans prepared to "settle in" and treat his feet, which had become cut and blistered by the last lap of the trek.

But rest and recuperation it was not to be. The enemy moved again, massively. Fertig's reports to Brisbane became very clipped—he was offering a minimum target for enemy radio direction-finder experts—but they were charged with trouble: the enemy's total strength in two columns moving inland was approximately that of a division: fifteen to eighteen thousand men.

On one of these critical mornings, Evans was counting off the minutes up to thirty: a half-hour's rest, then he would try his weight upon his bleeding feet once more. By that time there should be some deadening of the pain in response to the ampule of morphine he had taken upon awakening that morning, the fourth since they had adopted their fugitive existence to save their lives. Feet that had been in poor shape to begin with had responded in the only way they could be expected to through four days of wading in stream beds and grinding along the rockiest trails they could find in order to leave undetectable signs of their passing. Fortunately the other members of his little safari, Filipinos supplied by Wilson, were in better shape so they had been able to relieve him of most of his pack. Evans examined his feet. The abrasive action of sand and gravel had planed off all normal calluses; there was a constant oozing of blood from exposed capillaries. But for a while the pain would be less. And they had to go; the pursuing enemy was never far behind.

On June 24 Evans set up the ATR4A and hopefully called into McClish's net. Far from receiving encouragement, the wonder was that he made contact at all. At that moment McClish's headquarters in northeast Mindanao was being immobilized by enemy shelling. Pianing had been captured. There was no news of Fertig—but maybe no news was good news, for certainly the Japanese would have made much of it had they taken him. Evans shut off the receiver and packed it. But they would have to boil some drinking water before they could go on. While he was chopping he struck his left big toe. The log had been rotten inside; the ax had driven through.

The morphine partially blocked that pain, too. Evans bound the lacerated member and off they went. Evans trailed and the others helped, for he seemed to be like a man in a dream state. Probably he was. At noon a runner caught up with them with the news that an advance patrol of the enemy was drawing closer. There could be no more rests for a long time.

The next days were torture. He felt that it was not only too dangerous for him to risk the narcotizing effect of the morphine, but unfair to the others who had to be alert for every possibility. He took no more. The pain was so great that it seemed to numb his brain. They went on, trying to escape via the Sibagat River. The banks became almost cliff-like and the current in the constricted gorge tore at them. From the banks, boulders weighing tons had fallen long ago. They were smooth and slippery through the action of the rushing water. The men clung to them like wet insects and inched their way forward. Evans sent a scout ahead. The rest disposed themselves to cover the river to the rear. Through a haze of exhaustion they waited. Then the scout came back. That night they would sleep and sleep, proclaimed Evans, for there was a friendly village nearby.

At ten o'clock, when they had just dropped into the dead slumber of beaten men, a runner slipped into the village. The enemy, he said, had done a forced march and already was "a mile and half away only, sir—at Afga."

Evans fought sleep and concentrated on what the man was telling him. Afga, he had said. But Afga was in *front* of them!

They held a quick council. The enemy was back of them, too. How far back? That was the question, and on it would depend their freedom, probably their lives. The river that had served them as an escape route thus far, however brutally, now was their trap. There was only one hope—to scale one of the banks, then climb a mountain and drop down the other side of it into a valley that ran parallel to this one. If they could reach it, they should find an old mining camp. At the camp they might find Major Vincent Zapanta, one of McClish's men. But to reach the point where they might scale the bank and start the climb, they would have to go back along that hellish river bed with its clawing current and its slippery black boulders, and this time they would have to do it at night.

But one more gamble they would have to take: they could not push on without some sleep; it would be sheer guessing as to how long they dared sleep before one enemy or the other, or both, would close in on them.

At three o'clock the sentry shook them awake. Soon they were slogging in that Mindanaoan valley of the shadow once more. Scouts declared them to be safe from immediate detection. Accordingly, they lit their way with the red flames of nipa rushlights. All night they stumbled forward, driven by an urge to live that was stronger than the drive simply to lie down and await the enemy. With the coming of daylight each could see the marks exhaustion had left upon his companions.

Then came the heat. If the night had been grueling, the time until noon was like acid and salt poured into open wounds. Without sleep, rest, or food, they went on—laden with arms and ammunition.

What happened next probably was due to the disorganizing effect of exhaustion; perhaps there had been momentary black-outs for more than one of them. At any rate, Evans realized that he had called to José, his personal boy, about something or other and had received no reply. He stopped and called again. Still nothing. He turned back a few paces to speak to the others back of him—but there was no one back of him. He was alone.

For a long time, it seemed, he was stunned into complete inaction. Then the reflexes that had been driving him ahead for days took over and he found himself climbing again. Somewhere ahead was the mining camp, somewhere ahead he would surely find Vincent Zapanta.

The period that followed became mercifully analgesic. His mind began to play tricks on him. He was sure that he saw people up ahead. But they turned out to be rocks and trees and stumps.

Nevertheless, he was aware of speaking to them as if they were people.

Then there was no sound at all. It was quiet, deliciously quiet. Nor was he climbing any more. In fact, he was lying down in a hut of some kind.

His mind cleared and he remembered. He had reached the summit sometime in the late afternoon. On the ridge he had seen the tiny native hut, set high on the usual piles. It looked unoccupied. He had been glad of that because he was in Manobo country, and the Manobos were not hospitable to strangers—in fact, he had heard that they were cannibals. He had climbed up the shaky ladder. He doubtless had slept, but he did not think that it had been more than an hour. Before it got dark he had to take stock of his position. His body hurt in a hundred places at once as he tried to move. But he got down out of the hut.

Directly before it the mountain sheered off precipitously. He peered down the magnificent escarpment that must have been at least fifteen hundred feet high. He looked across the chasm. The country rolled away in ridge after ridge to the sea. Then he studied the valley below. Someplace down there Zapanta must be located— with food. But without more rest he could never descend the mountain to the Wawa River. He pulled himself back into the hut and slept.

He had no idea when it was—still that day or the next—that he awakened to find himself sitting up, staring into the haggard features of José, his boy.

Nothing seemed to surprise him any more. He asked about the others. The boy shook his head. Evans knew better than to ask whether José had any food. The Filipino was hardly in any better shape than Evans himself. It was agreed that they could go no farther without rest. In an instant they fell asleep.

Evans awoke with a sense that someone was climbing the ladder. He covered the trap door with his automatic, but the face that emerged was that of Ramón, another of their safari. The boy nearly fell from the ladder with fright before Evans shouted reassurance and dropped the gun. Ramón, looking comparatively refreshed, told them that the rest of the party had decided to try to escape by another route once they had become separated. He had asked permission to leave them and it had been granted. He had climbed the mountain alone, as had José, but for a specific reason: he knew that at the top was a hut that belonged to his brother-in-law. This was it. He planned to rest there, although he was not too tired, as he had slept the night before.

The night before? Yes, the boy replied, puzzled at Evans' question. After all, they had become separated two days ago.

Evans reached into the pocket of his filthy khakis. Ramón's face brightened at the two pesos. They would be his if he would but go down the mountain, find Zapanta, and tell him to bring food and water because he and José were too done in to move.

The boy left.

Sometime during the night he returned. Evans' eyes alternated between the chicken he carried and the companion he had brought with him. It was not Zapanta, but a Monobo native armed with a long spear. His hair was done up in a bun at the back of his head and on top of his head he wore an old hat of sorts, rather like an inverted gravy boat. His eyebrows were plucked to a thin straight line. Twin red streaks of betel-nut juice ran down the sides of his mouth.

Ramón's cousin, so Ramón said, spoke no English. But José suddenly broke into animated conversation with him. He was himself of Manobo extraction. The immediate point of the torrent of words seemed to be the chicken. And Evans noted with satisfaction that it changed hands forthwith.

Hardly waiting for the roasting to be completed, they tore it apart and wolfed it down.

Ramón and the odd character curled up on the floor. Evans slung his jungle hammock above them. His last thoughts were of an account of an incident in 1937, when Manobos had killed a number of people not far inland from the coast. Their heads never had been found. This spot, he recollected, was much more isolated than that location.

At daylight, sliding, clutching at bushes, slipping in reddish mud, they went down the mountain. At the bottom they rested, bruised, cut, and shaken. They resumed by wading in the rock-strewn bed of a small stream to conceal tracks. Then came another hour along the faintest of trails in the jungle. A halt. And suddenly Evans was aware that they had been joined silently by another spear-carrying native, long and lithe, as queer looking as the first. José told Evans that Zapanta was in a hut belonging to Datu Pataday. Evans had heard of him as one of the most energetic headhunters in Mindanao.

At the end of another hour they crossed a small stream. Before them was a native house of bamboo frame, split bamboo floor, and a nipa-thatched roof, the whole set on thin piles about six feet above the ground.

Peering inside, Evans drew a long breath of sheer thankfulness to see Zapanta and some of his boys. Against the opposite wall was a native more than six feet tall. José told him that this was Datu Pataday.

Evans surveyed his "host." The headhunter was attired in a native-woven shirt of abaci decorated with stripes and designs of bright colors. His shorts were made of the same material. On his arms were numerous bracelets of metal and stone. His mouth drooled red betel-nut juice and lime.

In one corner of the house were his aged mother and father, also drooling betel-nut juice and lime; in another were three young women attired in brilliantly dyed native cloth. Near the center of the floor lolled a native boy, obviously a congenital idiot. His chin was ropy with saliva. His mouth was twisted upward in a perpetual foolish grin. He groveled in the midst of his foul discharges. Near him sat another in the straddle-legged posture of the Mongolian idiot. Again and again he was tormented by his relatives, who slapped and pinched him. They were delighted at his whimperings of pain and dull resentment.

Despite his revulsion, Evans' exhaustion and the deplorable state of his feet ruled out any choice. He dropped on the floor and instantly was asleep.

For three days, between long, deathlike slumbers he was aware that Zapanta was endeavoring to convince the coldly listening Datu, who had never seen either a white man or a Japanese, that the paper money he carried actually was money, and that they wanted to pay for everything.

Zapanta had quietly arranged that one of their boys always remained awake, just in case.

By the third day the Datu was plainly disinclined to continue his role as host. Evans had been playing for time to allow his feet to heal. Now he became more concerned about his head. He was not happy that Datu's queer-looking sons—they of the sharp spears— were to be the guides. Once they were on the trail, he put them immediately before him. Eventually they came to another untenanted hut. That night they ate a poor meal of dried corn. Tired and still hungry, they went to sleep.

All the next day they pursued their trail. Roasted green bananas were their lunch. Evans rated them as being like a mixture of cotton and uncooked com meal. But that was all there was.

Late in the day there was a conference between the guides. Once more they must be close to Sibagat and the Japanese. But they had to take that chance: after all the enemy was all around. (Although they did not know it then, a patrol sent to capture them had passed just on the other side of the stream from Datu's house while Evans had slept.) But if ever they were to reach Fertig's headquarters again, they must break through somehow.

With only a warning hiss, the guides suddenly leaped into the bush. Evans followed, flagging the others behind him.

A patrol was coming toward them. He saw a Japanese helmet. His automatic covered it, but Zapanta shouted: "Don't shoot, Doc! Jesus, Doc, it's friends!"

The helmet had been taken from a dead Japanese. Together the two parties made for the barrio of Sibagat, in turn frightening the wits out of another little band of refugees who had taken shelter there. The Japanese were known to be very near.

That night a Filipino boy raced into the camp.

"Hapon!" he cried, jerking his arm backward. *"Hapon!"*

There was barely time for them to seize their weapons before the firing broke out. It was every man for himself, no man knowing where to leap to save his life. Evans and Zapanta never knew the fate of the others, but in some miraculous fashion they had not only leaped together, but in the right direction.

If they had saved their lives by their accidentally proper actions,

they soon realized that perhaps they had only postponed their end, for now they were thoroughly lost. Except for their packs and guns, they had nothing. The two decided to sit the night out. Through the hours they fought mosquitoes and tried to avoid centipedes, whose bites could cripple.

The next day they wandered, trying to get back to some landmark they could recognize. They never did. But they did come upon another village, this time all Manobo.

As near as could be determined later, this must have been early July of 1944. From then until the middle of October these two, an American physician and a Filipino who once had been a *maître d'hôtel* at one of San Francisco's finest, lived an incredible existence as adopted members of one of the most primitive tribes encompassed within the boundaries of any civilized country. Using his knowledge of herbs and the few pills of various kinds that he still had in his pack, Evans set up another "clinic." Zapanta was his partner. The Manobos trusted them and allotted them a large nipa house for their activities. In return for Zapanta's carbine, and instructions in how to use it, the natives gave the two men a "percentage" of what game they were able to kill with it.

One of the Manobos knew a smattering of English. There ensued prolonged educational sessions. But it was almost beyond the resources of Zapanta and Evans to explain something as complicated as an automobile to these people who found even the simple wheel a marvel, or to tell why white men and Japanese were fighting each other when both lived so far away across seas that any man could see were much too immense to be crossed. The Manobos could see Japanese aircraft flying above them, yet to include this into some comprehensive scheme of social relationship, good or bad, was something so impossible that it was wisest simply to ignore the fact of the aircraft altogether.

In August they could hear restless mutterings from the east. They did not know it then, but these were the first bombings of the Davao area by American air. (And deadly accurate, too, thanks to UU2 and the others in Charlie Smith's wake.)

On the morning of September 9 Evans' boy shook him awake. From long experience, Evans' first thought was that this was another attack. But the boy was shouting for him to come outside. There were many airplanes now; he held up ten fingers many times. Evans rushed out.

American bombers. And there were more than sixty of them!

There was a big celebration in the village, the Manobos not knowing quite why, but joining in with zest. Zapanta and Evans got somewhat drunk on tuba juice.

Then deliverance. The planes ultimately found the enemy columns and blasted them under.

Bidding their "brother tribesmen" farewell, Zapanta and Evans got into a baroto and made for where they thought they might find Fertig—if he lived.

He did.

On November 23, 1943, I had gripped Evans' hand there in the pre-dawn heat of Darwin, wondering what might happen to him. It was almost fourteen months to the day that I shook it again, in Brisbane—and learned from his own lips what *had* happened to him.

The Charlie Smith Way

WHILE ADVENTURE IN THE RAW HAD BEEN piling up unremittingly to liven and oftentimes threaten the life of the "Mindanao Mender" in the first half of 1944, Charlie Smith had been making history in his own way considerably farther to the north. It will be recalled that this was Smith's second penetration into the enemy's back yard. His first had been as a member of the old "Fifty" party. On that occasion he had sent his first message from his cozy hide just above Davao Harbor, one of the true enemy strong points in the south. It had become difficult for him to obtain food because of the enemy in large numbers "practically on my front doorstep." He decided to establish some degree of equality. One of his early messages had read:

> FIVE THOUSAND-TON WHITE CARGO STEAMSHIP AT POSITION FIVE-TWENTY NORTH ONE TWENTY-FIVE THIRTY EAST EIGHTEEN HUNDRED HOURS TWELVE KNOTS. THIS SHIP HAULS RICE TWICE WEEKLY COTABATO-DAVAO.

The "fix" was relayed to Perth at once. Perth pumped it north. The white cargo vessel forthwith ceased to haul this staple of the enemy's diet to Davao, or any place else.

Leaving UU2 in capable hands—so capable that ultimately Fifth Air Force bombers would score one of their outstanding successes in the Davao area—Smith was ordered to make his way back to Fertig's headquarters in time to join Chic Parsons on the frightful trek with the American escapees to the submarine rendezvous on the south coast of Mindanao. He allowed himself little recuperation time in Australia but fell to with characteristic generosity in preparing the newest up-bound party to go in November 1943. It was at this time that pressure developed for establishing the eastern anchor to the trans-Philippines belt of agents and watchers to accomplish two main objectives: to provide a complete screen for reporting enemy movements from Luzon southward; and to set up an efficient "underground railroad" for the movement of agents and guerrillas between Luzon and the central Philippines, thus opening up Luzon internally. Under Whitney's characteristically energetic direction, things went forward at a rapid clip. The only factor undecided was the selection of a leader. He spoke to Smith. As before, there was no hesitation. Whitney emphasized the extreme hazards of such a venture, which meant eventually penetrating all of the strongly-held San Bernardino Straits area and even southern Luzon itself. Smith's reply was undramatic, completely selfless: "It's my old stamping ground; I know it like my own hand. You wouldn't think of sending anyone else, would you, sir?"

It was after an anxious period of silence that the Bureau had its first indication that the little Masbate mining engineer and his party had established "squatter's rights" in enemy-held territory on Samar. The night of December 20, 1944, carried his wireless note, clean and sharp: "MACA" three times repeated. In the G2 office the successor to the late Colonel Merle-Smith made an observation that was as prophetic as it was original. Lieutenant Colonel Archie McVittie, who formerly had been associated with Merle-Smith in New York, read the brief message form. "And that, gentlemen," he said, "is General MacArthur's signature on the death warrant of the Japanese in the Philippines."

Charlie did not stay long on Samar. It was, he explained, not close enough to the enemy. Across the San Bernardino Straits was the Bondoc Peninsula, and Legaspi, where the enemy had made one of the earliest invasion landings of the war. The enemy was there—that was where Charlie would be. He set up a sub-station on the peninsula although the MACA net control was continued on Samar. He initiated expansion operations. To one of Fertig's best radiomen, loaned to him by the bearded Mindanao boss, Smith said:

"We've got to get closer to Manila; that's where the enemy is."

"There's a few of them around here, too, sir," replied Captain Robert V. Ball, one of Charlie's own kind.

Smith left Warrant Officer Stahl there and sent Ball to Baler in the

Bondoc area. It was another step toward the enemy's heartland.

Ball went, and the move proved to be one of the most significant for AIB and troublous for the enemy of any made by the cagey little Smith. In May 1944 the local radio link between the two points along the San Bernardino coast came in. Hardly had the net begun to function when KAZ picked up a "hot" one from MACA. It was directed for my urgent attention.

A guerrilla courier had broken through to Stahl's position on Samar. He claimed that he had come from Luzon. He also claimed that his chief was one "Andy" (Bernard) Anderson, an escapee from Bataan's hell, who was exterminating the enemy wherever he found them with his compact guerrilla band northeast of Manila. Through the grapevine "Andy" had heard of Allied Intelligence Bureau and that I was connected with it. He wanted help, wanted money, arms, drugs. In return, he could supply vitally-needed information.

"Andy" Anderson! Could it be the Andy of whom I had once been a messmate at Selfridge Field, and in whose eyes I had read uncomplaining acceptance of his own fate, one night in March of 1942 when I had said good-by to him on Bataan Peninsula? I had been evacuated; he had remained. Only *that* man could answer certain questions that could be put to him by radio. The answers, if correct, also could serve as the keys to a cipher system. Ten questions went north by night schedule. They were such as: "What was the hobby of the cook in the BOQ where we bunked in 1940-41?" He would have to answer: "woodcarving." "What was the name of the pet monkey at a certain hidden mess at the edge of Bataan Field?" He would have to answer: "Tojo." And so on. The next night answers came back, correct to the last word.

We had found a strong arm in vital Luzon. Plans were laid immediately by PRS to make a submarine contact for supplying these people. There were others as a result of Smith's "underground railroad," such as Major Robert Lapham and Lieutenant Colonel Russell Volckmann.

We wondered how word of the Bureau had penetrated so far so soon— even to the details of headquarters personnel; in some ways it was disquieting. It would not be until after the war that Andy would reveal that it had occurred as early as 1943 through one of Villamor's ever-spreading tentacles. "No one seemed to know for sure where *he* was, but his men certainly showed up in the most unexpected places," Anderson said, in what he probably had no idea was one of the best tributes to "Planet." He added that W-lO's men also had contacted other noted guerrilla leaders, such as Lieutenant Colonel Merril in Zambales and Lieutenant Colonel Ramsey, just outside Manila.

But this is Charlie Smith's story, for had it not been for him the San Bernardino corridor might never have come about and the "softening up" of Luzon might have been delayed indefinitely.

The time came for Smith to smuggle radio equipment up to a position in the vicinity of the Polillo Islands, east of Luzon. The instructions whistled through the air. Smith's "Roger" came back from the Bondoc. That was all, except that he was readying a small launch for the smuggling act. He was nearly prepared, but there was one annoying insufficiency. Smith decided to solve it his own way. He sent a message that should take its place among the immortals of warfare. It said:

DEPARTURE DELAYED OWING SHORTAGE OF LUBRICATING OIL. HAVE ARRANGED TO GET SAME FROM NIPS.

The Japanese were quite aware of Charlie and were determined to destroy him. They sent numerous raiding parties into the Bicol with instructions to block off all escape routes for MACA and take him and his organization dead or alive. MACA went off the air, and with it nearly every station in the net. The enemy was closing in.

There was only one thing wrong with the Japanese concept: It presupposed that an avenue of retreat naturally meant retreat upon one's own lines. From a remote station as yet intact came the second of Smith's classic messages.

SITUATION HERE TOO HOT AM RETREATING TOWARD MANILA.

He did, and went to work again at a new address.

Japanese documents captured at the end of the war described the worries Charlie had provided. Said Watari Group headquarters, July 3, 1944:

The appearance of new enemy wireless stations is as frequent as before. In particular, the enemy has recently brought in many small-type sets which are used for communication within the islands. Wireless communication has increased considerably. The bandits (sic!) seem to have learned about our interception by plotting, for they are skillfully concealing their stations and are successfully preventing interruption of communication by our punitive units. Station MACA alone is equipped with 10 to 20 wireless sets. More accurate information must be collected to destroy these stations.

Following the determined raids made by the enemy parties came another report (likewise subsequently captured and translated) proving that it was one thing to know about Charlie and quite another to get to know Charlie. It said:

> Enemy communication continues brisk, and he seems to have made considerable preparations against interruption by our punitive operations. Although we captured seven transmitters and fifteen receivers in operations against Station MACA on Samar, communication continues as before, interrupted only two weeks.

Of course, there would have been for Charlie, as for any of the others in the Bureau's chancy business of spying on the enemy, the same frightful consequences of a single misstep, but his operations always seemed to be characterized by a certain impishness.

The situation had become quite uncomfortable for him as a result of enemy action in the Bondoc and he considered it necessary to move again—farther north. This time he planned to scout the area ahead by dispatching a messenger before him in a small launch. But there hardly was fuel for one. Worse yet, the enemy which had provided him with oil on the previous occasion apparently was suffering as severely as he was from a lack of gasoline. Charlie was reported to have murmured "Heaven will provide," and started the messenger off. It was after the war that Charlie told me the rest of it, his eyes merry and a hint of a smile under his mustache.

"The man watched the fuel gauge drop toward zero and he imagined sad things, so sad that they made him want to weep and thus he almost missed the first of two fifty-five-gallon steel drums floating by—probably from a torpedoed ship up north someplace. Well, he snared it aboard, figured the currents, and just let the other one drift southward toward me, thinking that I'd likely know enough to take it in. Then he went on."

Charlie was asked if he had encountered the second drum in that wide ocean.

"Oh, sure," he replied simply.

Smith has got some medals tucked away somewhere, high decorations that go only to those whom the country believes to have done outstandingly brave and valuable deeds. Yet probably the finest tribute to him was paid by the warring guerrilla factions in the east-central

Philippines. They were desperate men and leaders, or they never would have survived. They came to Smith with a proposition: If Charlie would agree to be their boss, they would all forget their personal differences and ambitions and come in as one force.

They awaited his decision. Hard, grim faces, intense eyes, lean jaws with blue stubble on the sunken cheeks—leather-skinned hands that never strayed more than a few inches from ready carbines. ...

Two weeks later the authorization from Brisbane arrived. Now it was "Lieutenant Colonel Charles Smith, the official commanding officer of the Samar Area."

When the tremendous forces of the combined army, navy, and air eventually smashed in and began the pincer move up from the south and down from the north on Manila, Smith was called to the field GHQ. His radio net was made available to General Walter Krueger, as well as the file of reports of Smith's intelligence agents operating into Ball's station on Baler. The hard-headed field commander realized from this and other data that he was opposed by a far larger force of Japanese on Luzon than he had believed. Wisely, he held two divisions back for the fight he knew would come around the Baguio area. He resisted being drawn too far south too soon with too few men. The move was destined to become a highly personal matter for little Charlie Smith. He later spoke of it this way:

"General Krueger's Sixth Army didn't come down but I did. I was in front of Sixth Army doing some reconnaissance for General Krueger's G2. I was also in front of the Japanese Army. I was between 'em. They didn't call it 'No-Man's Land' in this war. But I was in it just the same—all by myself. I didn't sleep very well," he said simply. "Funny the fixes a man will get himself in, ain't it?"

The Captured Plans

THE HEAVY LAND CONTESTS in which Charlie Smith had been caught broke the back of Japanese power in Luzon earlier in 1945 than had been anticipated by the Allied High Command. These military victories were the consequence of successive catastrophes that had overtaken Japanese imperial naval strength in the battles of Leyte Gulf late in 1944. The Japanese Navy literally ceased to exist then, and its destruction isolated Japanese land strength in the Visayas and Luzon and enabled MacArthur to drive ahead relatively free from sea and air threat. Schedule after schedule became obsolete before it had even been implemented. The way was opened for the attacks on the enemy home islands and the end of the war.

The Philippine network established by AIB played a vital part in insuring our success in those tremendous naval engagements. The Allied naval command not only went into those battles with the broad Japanese naval strategical concept outlined before them, but they knew what enemy ships were likely to participate, what were their fuel ranges, their fire powers, their vulnerabilities—even the names of their commanders and some of their personal characteristics; all this and more, when prior to that time the utter paucity of information about the "mystery navy" of Nippon had been a matter of gravest concern. Possession of the Japanese "Z" plan together with the wealth of sustaining data enabled exact tactical planning, the matching of weakness with strength and strength with greater strength.

How had this come about?

In the southern Philippines the day of March 31, 1944, had been particularly heavy with the humid heat so characteristic of the region. As the breathless afternoon went into the brief dusk of the tropics, a light offshore wind made it a little more comfortable, especially in the high country inland and to the south of the city of Cebu, the most important city outside Manila. The enemy was in strength there and James M. Cushing's Cebu Area Command guerrillas gave the city wide berth, preferring to move like the shadows they were in the mountains to the west and south. Lieutenant Colonel Cushing endeavored to keep them deployed, so that no sudden enemy strike could effect more than local damage, if any. There was only the usual handful of commandos at his headquarters in the mountain retreat of Tupaz. Cushing particularly welcomed the relief from the heat, for he was a sick man and so debilitated by recurrent attacks of malaria that he could get about only with the aid of crutches. He looked off to the east and saw a massive build-up of towering black cumulus clouds flickering with a continuous play of lightning. Like gunfire, he mused.

That thought gave rise to another. If what "they" had been saying was correct, there had been some highly significant gunfire off there to the east—far off, in the Palau Islands, to be exact. "They" were the voices that came to them through the medium of radio receivers deeply hidden at the edges of some of the barrios. Only greatly trusted individuals knew of them and were allowed to listen by means of the tightly cupped earphones. Through them, when the static was not too bad, they could get the bulletins out of KGEI in San Francisco. There had been exciting news. Powerful American naval units had been sweeping westward across the central Pacific to rendezvous with other elements coming up from the now well-contained Solomons. The combined strength, including eleven huge carriers, spearheaded by heavy air attacks from their carrier aircraft and land bombers from SWPA, apparently had dislocated Japanese naval operations, especially around the Palaus, the stronghold of Admiral Mineichi Koga, the commander in chief of the Combined Imperial Fleet. Neither KGEI nor Cushing could guess the true extent of that dislocation, or that at that very moment momentous events were occurring in consequence of it, events that would deeply concern him. Cushing observed the impressive storm front again. It would be well out to sea, probably, and it would be stiflingly hot down in the coastal regions.

In the barrio of Balud, San Fernando, just inland a few hundred yards from the eastern shore line of Cebu, a handful of Filipinos in nondescript clothing, mostly tattered and patched khaki, lolled about or squatted on lean haunches. Their black eyes were on the deserted provincial road that led away from them down the coast toward a point where there was a Japanese detachment that sometimes raided along the road and drove the fisherfolk and farmers into the mountains. The Filipinos were unmoving, for the evening was breathlessly hot. But they were careful not to abandon their weapons even for a minute. Only two had guns. The others were armed with curved bolo knives that gleamed dully in the fading light, or with pikes, some of which had knives on the ends. They were volunteer guards.

The evening went into night. Still the lightning played around the cloud mountains on the sea horizon. The storm was moving farther away. There was a moon. It went down early.

It was then that something happened out at sea.

First it was a sudden bloom of light. But not from the storm. And it was red and yellow instead of the blue-white of lightning. The guards were all alert now, straining eyes to seaward. The light swelled and pulsated. It grew bright and they could see silhouetted against it the little black shapes of fishing boats moving out toward the source of it.

It was, the guards surmised correctly, an aircraft crashed into the sea and burning. In no time the half-dozen guards were joined by twenty—fifty—a hundred others, all armed one way or another. The air was alive with the babble of their excited voices.

It took the fishermen a long time to reach the burning airplane, and more time for them to come back to the beach. The babble increased to a din. When some order came out of the confusion it was apparent that the village of Balud was host to ten Japanese, most of them so injured as to be unable to more than move. Only one of them was able to stand. Even practically naked as he was, and badly hurt as well, he gave an appearance of command. He regarded the menacing circle unafraid, even once indicating to one of the men with a gun that he should use it upon them, the captives. There were many cries of agreement.

But the cooler heads among the guards prevailed. These men, whoever they were, could be made to talk if they lived. Cushing's guerrilla soldiers in the mountains would know how to question them.

But the Japanese could not walk. Even the leader now sagged. Very well, they would be carried in improvised litters. It was then, as the queer safari started into the dense country inland, that the waterproof container which had been in the possession of the authoritative-looking Japanese was produced. This, too, would go to the nearest guerrilla unit.

After an exhausting journey that lasted until early in the morning, the procession reached an outpost of the Eighty-seventh Regiment of the Cebu Area Command. From here runners were dispatched, both to the regimental headquarters and to Cushing's headquarters at Tupaz. At one of these points was located the little ATR4A that Villamor had spared Cushing from his stock. Through the medium of this precious set a signal went out to be snared by both 4E7 on Negros and WAT, Fertig's net control, on Mindanao. It told of the capture of a whole case of important enemy documents including what seemed to be a Japanese cipher system together with ten men, some of whom might be of high rank. WAT relayed to Australia.

At Heindorf House and GHQ the message created a tremendous stir. Immediate steps were taken by the Navy to divert an operational submarine to Villamor's old site on Negros to pick up the party. A little later a message came from Cushing himself, stating approximately: "Ten captured Japanese coming this headquarters; believe from Palau; one thought high rank probably General Furomei commanding land and air forces Macassar. Advise action. Constant enemy pressure makes our position precarious."

It was not long after the receipt of this message that another came which created an even greater excitement. It intimated that the party included no less than the commander in chief of the Combined Fleet. That would be Admiral Koga himself!

It would be truly the catch of the war.

The messages were the first of a series that formed the basis for confusion that would prevail until long after the end of the war. In fact, to the end of their lives there would be good, honest men who could never believe anything other than that they had been involved in the taking of Admiral Mineichi Koga. They had excellent reasons for believing so, including, among other things, the statement of the impressive captive himself that he was, in fact, Admiral Koga.

This statement was made later in the trek, when eventually the party had been carried every step of the way over treacherous mountains to Cushing's compound at Tupaz. Here they were given the best he had in medical care. The party had refused to acknowledge that anyone in it could understand any Filipino dialect or English. But one of Cushing's regimental interrogators knew some Japanese as did Cushing himself. That broke it, and now the leader admitted that he spoke fluent English, as, indeed, did most of the others. (For some reason, however, Cushing still addressed him as "General" although postwar records failed to disclose the name of Furomei or such a post as "commander of the land and air forces, Macassar.") Another ranking member of the party was Commander Yamamoto.

By this time, however, Cushing knew something else all too accurately: he knew that he never could successfully convoy his captives to Negros for embarkation on the submarine that was coming, for now the Japanese on Cebu were aware of the captures, and they were sweeping through the helpless areas of southern Cebu like avenging scourges to recover the prisoners. Cushing sent an agonized message to Australia. The answer made the sweat break out on him. It was:

ENEMY PRISONERS MUST BE HELD AT ALL COSTS.

He stared at the message and slumped into the sawed-off box that was his office chair. Surely General MacArthur could not mean that. He himself had declared Manila an open city to save the civilians from enemy rage when he had retreated from Manila to Corregidor and Bataan. Now Lieutenant Colonel Homisi, the Japanese commander in Cebu, was acting on orders from an adamant Tokyo that the captives and anything taken with them would be recovered at all costs. Homisi knew that only harakiri could atone for his failure, and he did not intend to fail, even if it meant the extinction of every living thing in southern Cebu. Soon the country was under a glare by night and a shroud by day as men, women, and children alike died in their ravaged barrios.

Cushing knew that it was impossible to comply with that GHQ order. His forces were being scattered. The enemy shipped in a whole unit of elite marine raiders. Cushing had only twenty-five commandos with him. He would use two of his best men to spirit the documents to Negros, but that was the best he could do.

These men were joined by another and together they did accomplish an incredible infiltration across Cebu, across the water separating Cebu from Negros, and on to Villamor's old site, now commanded by Lieutenant Colonel Andrews. Always one step ahead of pursuing Japanese the fleeing trio went, arriving at the rendezvous only as the submarine came to the surface, took the case aboard, dropped back into the depths, and pointed its sharp bow for Australia at top speed. Then the raiders swept over the Negros headquarters site. But they did not capture the exhausted, triumphant men.

In Brisbane, Cushing's newest message reiterating that he would be compelled to surrender his captives to spare the civilian population generated wrath in the office of the Commander in Chief. He demanded that Whitney send a message announcing Cushing's immediate discharge in disgrace—and probably worse when our troops returned to power.

In the back room Whitney paced in short, jerky steps, his pale blue eyes blinking back the intensity of his feelings.

"Cushing couldn't help it. He couldn't *help* it!" he muttered again and again. "It meant the blood of thousands of helpless Filipinos on his hands."

He seized his uniform cap and went to GHQ. He was gone half an hour. When he came back, he flung his cap into a chair. "I've got twenty-four hours to prepare a defense," he announced shortly. He ran everyone out and locked the door. He did not go to bed at all that night as he called upon every resource in a legal armory gained in twenty-two years of practicing law to prepare that defense.

On Cebu the situation was becoming desperate. Scared, exhausted runners were arriving at Tupaz at intervals to back up radio reports that the powerful enemy had routed the guerrillas in every direction. Cushing knew it would be only a matter of time before they would be overwhelmed. But he would play for every minute of that time he could get. On a neighboring hill he had prepared a final defense perimeter at Maserela. It was not much but it was all they had left after Tupaz. He gave order for the evacuation, should there be an assault on the compound.

At about three o'clock in the morning it came. A crash of shots from the northeast told him that an enemy force from Cebu City had broken through their outposts. It was every man for himself to get to the perimeter *but* the prisoners must be brought along and secured there.

Cushing's custodian of records, a commando sergeant named Alfredo Marigomen, together with another commando named Meliten Endagan, helped the man they believed was Admiral Koga onto his litter and amid the flicker of small-arms fire, managed to evacuate the shambles of the compound and gain the heavy foliage of the mountain. Commander Yamamoto was carried by Commando Pedro Gabriel. The going in the dark was of the heaviest. The men panted, then gasped for breath. The admiral asked for water and Marigomen gave him some in a coconut shell. They rested, then stumbled on. Sometimes they fell into holes. Sometimes they had to pass the litters over their heads to by-pass the tangle of creepers and vines.

It was noon the next day when they all arrived at the perimeter of Maserela. Cushing was there. He told them that the old headquarters site had been razed and that the enemy in overwhelming numbers was in full control of the area except for their own small perimeter. He indicated that he wanted to talk to the admiral alone.

For a long time they spoke. Then the Japanese leader asked for writing materials and a messenger. He was going to propose to Lieutenant Colonel Homisi that he refrain from all further terrorism or attacks upon the guerrillas in exchange for the prisoners themselves. He signed the document. Cushing's hand shook so that he had to make two tries in signing his own name.

The messenger took the paper and together with a captured Japanese airman Cushing had been holding, disappeared down a gully where the still-warm bodies of many Japanese and Filipinos lay in the sun. For Cushing it was an eternity of waiting, for he knew that he had shot his last bolt: he had permitted the enemy to organize in strength along his pitifully weak front and he had offered to trade with that enemy in direct defiance of orders from the American Commander in Chief. The sickness welled up inside him and it was only through a haze that he saw the messengers slowly climbing back up the hill. One of them bore a paper signed by Homisi, giving the promises that he had asked.

Now came a scene unparalleled in the war to this time. Cushing shook hands with his captives. Headed by the admiral on his stretcher and accompanied by an unarmed platoon of Cushing's men, the Japanese prisoners filed slowly down the hill toward a banyan tree. From the bottom of the gully another platoon, this time of unarmed Japanese soldiers, toiled under the hot sun toward the tree. At a distance they halted and eyed each other. Except for the subdued clicking and hissing of insects in the afternoon heat there was no sound. Cushing knew that hundreds of enemy with their weapons covering every foot of his position and the platoon out in front waited, fingers on triggers, for one small hitch—or even for a word of command, should this prove to be a trap, after all.

There was no hitch. It was not a trap.

The two groups moved closer—merged.

Some of the Japanese soldiers held out cigarettes. The ragged Filipinos looked at them hungrily, then reached out. Little mists of smoke emerged from the circle as the men smoked silently and stared at the well-fed, stoutly-clothed soldiers before them. The stretchers changed hands. The two groups surveyed each other for another moment, then turned and retraced their steps.

It was done.

For three days there was silence and peace in the area.

No one was surprised when eventually the air beat to the crude air-raid alarms of hammers on empty shell cases. The planes came and the bombs crashed down. The war was on again, but this time against the guerrillas only.

The resumption of active war coincided generally with the arrival in Manila of the one-time captives. Promptly they were sent to separate hospitals as isolation patients who were to remain absolutely silent about their experiences. It would not be known until after the war that this was one of the devices by which the Tokyo government endeavored—successfully, as it turned out—to keep secret for the time being the fact that Admiral Mineichi Koga had, in fact, disappeared the night of March 31, and was presumed dead. It was to avoid risking a plunge of public morale at this critical time in the war that the Japanese Government had decided to keep secret the disappearance of such a distinguished naval leader, especially until all hope of finding him alive had expired.

Admiral Koga had indeed met his death that night.

Who, then, was the Admiral "Koga" that had been held by Cushing's guerrillas for a matter of ten days in 1944?

In the years following the close of hostilities the evidence slowly was accumulated. Much of it was based upon interrogations, included repeated questioning of the individual whom Cushing's men were confident, and remained confident, was Koga. Unless the years to come bring to light some even more dramatic and irrefutable evidence, it would seem certain that this captive was not Koga, but his brilliant chief of staff and second-in-command, Vice-Admiral Shegeru Fukudome.

Postwar researches have yielded the fact that late in February 1944 Koga had attended a top-level conference in Tokyo to put the finishing touches on his "Z" plan. This provided for the concentration of nearly the entire naval and air strength remaining to Japan—a very considerable strength indeed in every arm except carrier-borne air—for a sudden, all-out attack upon the Allied naval strength coming westward across the Central Pacific. Koga was ultra-aggressive. He proposed to offer tempting bait to the major American units to lure them within range of his air, which then would make a mass, even suicidal attack to reduce their effectiveness and make them vulnerable to the next phase. This would be a sudden devastating gun-and-torpedo smash calculated to drive Allied naval strength to the bottom and end for all time the threat to Japanese home waters. In Nippon's brief, brilliant naval history there was ample precedence for this: the destruction of the Russian fleet in 1904; Pearl Harbor in December 1941; the blasting of four fine Allied heavy cruisers in only a few minutes of a firelit night off Savo Island in the Solomons in August 1942. ...

Koga was to get everything; even land-based air units from as far away as Singapore were called up for operation out of Mindanao.

Back in the Palaus, however, he was disturbed by the tremendous aggressiveness of the Americans themselves. His own headquarters –the entire fleet anchorage area–came under destructive air bombardment. Then in a final brilliant exploit that got little attention in America but won the grudging admiration of the Japanese, American naval airmen thoroughly mined the very waters Koga regarded as "the keystone of Japan's inner defense zone." The Palaus rapidly became useless to him as a base from which to operate or direct his showdown battle.

Very well, then, he would direct it from a land headquarters– Mindanao.

He called up four-engined Kawanishi flying boats. He would take one, Fukudome the other. Since he knew by heart every detail of the Z plan and all the vast paper work that had gone into coordinating and implementing the concentration for the batde of the giants, he directed Fukudome to take the bulging case containing all the vital documents pertaining thereto, together with the cipher system applying to much of it. In the middle of the newest night air-raid alarm he took off. Fukudome waited only to watch the heavy Kawanishi swing into the southwest before he boarded his own craft, together with Commander Yamamoto of his staff.

That was the last Fukudome or any other man associated with Admiral Koga saw of him. Koga had gone to fulfill his announced destiny, for in his communications reference the Z Plan, he had written that he would fight it out on that line "–until the death."

Apparently death came to him and to all others aboard the first Kawanishi in that tremendous build-up of lightning-shot weather off the Philippines that night.

Fukudome subsequently stated that his own aircraft was unable to cope with the huge front, and to avoid it, swung far to the north. The four engines carried her high, so high that the lack of oxygen apparently affected all aboard and may have impaired the judgment of both the navigator and the pilot. At any rate, when it became apparent that they would have to put down someplace to replenish fuel consumed in forcing the engines, there was doubt as to their whereabouts. At last it was decided that they were off Bohol or Cebu. They believed that their chances of refueling on Cebu would be good despite the surprising amount of "bandit" activity prevailing on that island. They descended. Just when the pilot was attempting to get bearings for a final letdown, the moon itself went down and left the country below them in blackness. Fukudome pulled back on the control column to maintain buoyancy until they could con the situation some more. But the Kawanishi had lost her safe flying speed. She stalled, then sideslipped heavily into the sea. The remaining fuel had ignited. Apparently the promptings of training were such that even in this extremis he retained a grip upon the waterproof container with its vital documents. It was taken from him by the Visayan fishermen only after he had lost consciousness in the water.

In Brisbane the unbelievable array of documents was pounced upon by experts who reproduced every page of it. These reproductions were rushed to a day-and-night concentration of the best talent Colonel Mashbir's translators could produce until every word of it was available for the collators and the analysts at Naval Intelligence. Meanwhile, the originals were resealed in their waterproof container and rushed back toward the Philippines by submarine. It was hoped that they could be deposited in the sea in the area of the Kawanishi's crash to enable Japanese divers who had been assigned to sweep the ocean floor for them to "find" them. Thus Tokyo might never know for sure whether they had ever been in Allied hands.

But Tokyo was acutely aware of the fact that they were missing.

Another ultimatum had descended upon Cushing's headquarters, or what was left of it:

> ... RETURN UNCONDITIONALLY UNTIL NOON OF MAY THIRTIETH ALL DOCUMENTS, BAGS AND CLOTHING EITHER PICKED UP FROM THE SAID AIRPLANE OR ROBBED OF THE PASSENGERS.... WE NOTIFY YOU THAT IN CASE YOU FAIL TO FULFILL OUR DEMAND THAT IMPERIAL JAPANESE NAVY WILL RESORT TO DRASTICALLY SEVERE METHODS AGAINST YOU...

It was impossible for the submarine to reach the crash site by the expiration hour of the ultimatum. Cushing's message of June 16 was a model of official restraint:

PLANES BOMBED AND STRAFED CONTINUOUSLY FOR TWO WEEKS. SINCE THEN DAY AND NIGHT PLANE ACTIVITY PASSING OVER CEBU.

All that was a month before the clash of the naval giants actually occurred, yet the measure of the enemy had been taken.

Cushing would be there late in 1944 when the Americans swept in. He would meet them in his official capacity as commander of the Eighth Military District (Cebu) under MacArthur.

Whitney had won his plea.

Part 5
THE COMMANDOS

M.V. *Krait*, 1973. Restored in 1964, dedicated as a war memorial and used for training and recreation purposes by the Royal Volunteer Coastal Patrol. The vessel is now part of the Australian War Memorial's collection. Originally known as the *Kofuku Maru*, the Chinese built trawler was commandeered in 1942 and operated by the Services Reconnaissance Department, Allied Intelligence Bureau, for clandestine operations. The *Krait* was famous for its role in "Operation Jaywick" conducted by Z Special Unit, a successful effort to sabotage Japanese shipping in Singapore Harbor.

Limpets for Singapore

IN SEPTEMBER OF 1943 THE DIVERS OPERATIONS of the Bureau were extended over a front of thousands of miles. They were concerned mostly with the production of intelligence information and warnings of hostile intent. They were about to be extended to an even greater front, and this time the gathering of information would be subordinated to the primary objective of affecting sabotage. Villamor was still in the Philippines; in Brisbane preparations were being pushed for the insertion of the ill-fated Phillips' party; Read was recuperating in Australia; the silence from the Dutch party before Hollandia was ominous, and on New Britain, Wright and Figgis had been alerted to receive the "Gazelle Necklace" observers coming in by submarine. In the west, the Dutch had drained themselves white in their brave, tragic efforts to establish clandestine communications with the interior of Java. Singapore was even more distant and it was the enemy's western naval base. GHQ's cool reception of a daring plan for inflicting a blow upon the Japanese in their own stronghold at the tip of the Malay Peninsula was understandable.

Actually, the first admittedly vague suggestions for "Jaywick" project had been outlined to us quite early in the Bureau's history. The new unit was just hitting its stride when one Captain Ivor Lyon of the British forces asked for a confidential appointment. Captain Lyon wore the brass-buttoned uniform and rakish tam of the Gordon Highlanders. He was more recently of Malaya. It was that phase of his life which accounted for the cold, compelling look in his eyes and the taut, unsmiling mouth. Back there somewhere were his wife and daughter. The first stories said they were prisoners of the Japanese after the fall of Singapore. Then there were later hints that perhaps they no longer were alive. Lyon never referred to them. But to those in authority he spoke of schemes for getting back into Singapore Harbor which now, like Rabaul, had become a pivotal strong point and naval base for the enemy. If successful, his plans would not only supply GHQ with information of the area and the nearby East Indies, but would exact from the Japanese a very heavy price.

To negotiate hundreds of miles of hostile water unescorted before the target could be reached was an undertaking at which most hardy planners might look askance; to propose to blow up enemy ships within his own base by penetrating it in canoes sounded downright incredible. The men would be bearing devices known as magnetic limpets, charged with a high explosive.

It was obvious to Lyon that his proposals lacked the power to impress those at GHQ whose approval he would require. But AIB directed him to report to Lieutenant Colonel Mott. It will be recalled that Mott headed the Bureau's secret sabotage organization.

If Mott was terse, direct, and single-minded in his determination to hit back at the enemy who had run him out of Burma and Malaya, he was now confronted by his counterpart in the captain who gave him the British flat-handed salute. They measured each other coldly: Lyon believed he had found the right chief; Mott felt that in Lyon he had a perfect operative. Mott never doubted but that with the proper training, the right associates, and a potful of luck, project "Jaywick" could carry it off. But there was GHQ and its doubters.

"We'll simply have to provide them with a convincing demonstration," Mott said.

"Show how a limpet works?"

"I mean we must actually enter one of our own tightly guarded harbors and bug every last ship in it with limpets—sterile ones, of course," he added with a note of irritability that suggested he considered the point minor.

"I could do it, sir," said Lyon quietly.

"No, not you!" Mott replied. "You must be saved for the real show. There'll be the devil to pay when they discover we got it off—the leader will come in for some unpleasant publicity. Hummm." Mott thought for a moment. "Carey, of course. Who else?"

A few days later Captain F. W. Carey of the Australian Army was on his way north. On the coast of Queensland a good distance above

Townsville Mott's section maintained a training station. It was an isolated, tightly guarded place. Casual wanderers who might stroll toward it were intercepted before they got close and advised to take the air elsewhere. It was to this establishment that Carey repaired.

In the days that followed there were exercises in the sea just off the training station. They involved canoes, men in swimming trunks, and crablike objects that the swimmers clutched as they dropped below the surface. The metal arms of the objects were taped and painted gray. Between them was suspended a gray metal tube, slightly smaller than a quart milk bottle. The arms were in reality powerful magnets that caused the device to adhere with amazing tenacity to such things as the steel plates of a ship. Two limpets attached at different places on the hull would blast a five-thousand-ton ship, while three, expertly placed, could send a ten-thousand-tonner to the bottom. The exercises continued both day and night.

Townsville had become a boisterous garrison town, grossly overcrowded with troops of every arm and service in both Australian and American forces. To add to the congestion that boiled the length of Flinders Street were hundreds of naval ratings, for Townsville Harbor, once a sleepy, subtropical port of languid ease, had become a combination of naval-base and trans-Pacific unloading depot point. Merchant ships and transports of small tonnage and great were tied up or waited for their chance to get tied up. Escort naval vessels flying the Union Jack of the American Navy and the White Ensign of the Australian Navy were nested side by side while their crews roared ashore—except those on the watch.

There had to be a watch, for on one occasion enemy bombs had fallen not far away and on several other instances enemy submarines had been sighted off the coast. Still, after the tension of running the Pacific gantlet, any harbor seemed a snug harbor and it was likely that not all security regulations were as tightly enforced as they might have been. At any rate, even though later several seamen and officers recollected having seen during that particular night small craft, like bumboats, moving nonchalantly in the waters between ships, and some recalled hearing sharp, metallic clicks, like the bolt of a rifle being drawn, there had not seemed anything alarming in all this.

No one seemed quite sure who it was at dawn that spotted the first limpet gripping the hull of a ship. His cries soon were drowned out by the mounting scream of a siren. Then two more. Searchlights stabbed the pale dawn to examine a hull that might go asunder any moment. Shore police promptly bellowed for a blackout. Water police, shore police, military police—soldiers, sailors, and airmen appeared from nowhere and everywhere. All sought to advise, direct, and control concerning something they knew nothing about, except that apparently every major ship in the harbor had been mined by some secret device and the whole place was going up in a blaze of glory at any moment.

Safely away from the harbor area Carey changed into his normal Australian Military Force uniform. Meanwhile, his elusive limpeteers had rendezvoused at a secure hiding place. Carey made for town, thoroughly enjoying the unholy row along the water front, and went to sleep. The night had been strenuous.

Gradually the harbor authorities learned the truth about the stunt. They were advised to seek one Captain C. W. Carey, said to be taking his ease at a pub. There the concentrated might of military and civil law descended upon the sleeping man. Carey saw that they were in no mood for levity. He answered questions promptly and advised them to telephone AIB to confirm his claim of a "harmless practical exercise."

Unfortunately for Carey's immediate plight, AIB had not been advised by anyone of the plan. Mott was not to be found for some time, and, when finally cornered, was curiously vague as to the authorization for the "exercise" but most emphatic about the efficiency of the indicated results. Carey was dutifully meek, GHQ was properly exercised, and AIB was extremely active. The upshot was Carey's release on AIB's promise that he would be transferred to New Guinea. There were other transfers in and around Townsville Harbor that AIB had nothing to do with and many replacements appeared in officers concerned with the security of installations.

The Japanese apparently never heard about the incident, for if they had, they might have tightened their own security in Singapore Harbor.

Ivor Lyon and a team of the most carefully selected men "went underground" on a program of exacting training that included digesting all of the lessons learned in Carey's "Destruction of Townsville Harbor." For GHQ was convinced.

It was in the first days of September 1943 that "Jaywick" stirred. Off the west coast of Australia a small black craft, Diesel-powered, moved from her berth at the United States Navy's base at Exmouth Gulf, north of Perth.

HMAS *Krait* was really a stranger to these Australian waters. She, like her skipper, was a refugee from Japanese invasion. But unlike her skipper, she had once been Japanese-owned. It was from Singapore she had come; it was Singapore to which she proposed to return. She had been refitting at Exmouth for some weeks. Now as she moved out from under the high sides of U.S.S. *Chanticleer* she excited no particular curiosity. She was long and lean—seventy feet by eleven—and had a queer, high cabin rising just a bit aft of midships, giving onto an awning that carried to the stem. Inboard she carried a cargo inconsistent with her nonchalant exterior: canoes, limpets, tins of organic dye which could give the white man's skin the golden brown of the Malayan; and mounted in such a way that they were invisible from without, she carried some light machine guns. There was much more, details, mostly, for Lyon was a meticulous planner, an indefatigable trainer. This same trait would be manifest in the exact log he kept and the detailed supplemental record he would prepare.

By nightfall of September 2 *Krait* was rolling heavily in an unpleasant sea kicked up by a fresh southerly breeze. Despite the last-minute removal of some of her deck armor, *Krait* was not happy about her overloaded state and she would prove to be an uneasy craft in any kind of sea. Although they believed in her, the uncomplimentary remarks uttered by Lieutenant D. N. Davidson, the second-in-command, and shared by Lieutenant H. E. Carse, her navigator, were honestly descriptive of her caperings. The medicine kit of her doctor, Lieutenant R. Page, would come in for considerable use before the crew "salted down." Page was the son of Sir Walter Page, deputy governor of Australian New Guinea. Besides these and Lyon, *Krait* carried ten others, some crewmen, others operatives. Of various ranks were A. Crilley, R. C. Morris, J. P. McDowell, K. P. Cain, H. S. Young, W. G. Falls, A. W. Jones, A. W. Huston, F. W. Marsh, and M. Berryman. There was one compensation for the rough sailing: the wind was abaft and in four days and nights they should be high up toward the equator and the Lombok Straits through which they would have to pass before they could alter course to the northwest and sail through the enemy-controlled Java Sea toward Singapore itself.

The four days passed uneventfully, despite perfect visibility that greatly increased the chance of detection. On the night of the fourth, to the straining of her powerful Diesels as the *Krait* pushed against the swift current, they slipped through the narrow straits.

For two days they progressed as if they had been on a pleasure cruise. It was unbelievable. With only slight alterations occasionally they were able to avoid even remote contact with native fishermen. Surely their luck was tops. But they were taking no chances. With great care they applied the body dye that converted their tropical tans to the dusky shades that would allow them to pass as Malay fishermen. It would last for several days under the best of circumstances. Skin, hair, ears—never as youths had they washed their ears more thoroughly than they did now, but this time to dirty them. Even the nostrils and between the toes. Now the speed was more than six knots toward the Karimata Group of islands. Then, just before dawn of September 14, Carse shook Lyon awake and told him the sea was alive with junks and sampans. "We're right in the midst of them."

Lyon was on his feet and into his clothes. "Patrol boats?"

"Can't see."

From the dark wheelhouse Lyon did a quick survey. In the faint pre-dawn light he could see at least twenty small craft of every description. That there were others, he had no doubt, but he discerned only fishing craft.

"Drag that fishing net over the stem," he directed. "We've got to be one of them."

It worked. Gradually they were able to put distance between themselves and their industrious neighbors. Once they were hailed, but they ignored the cry, and kept going at slow speed.

They were free. For two more days their luck held. Now they were set for the Temiang Straits entrance to the Lingga Archipelago. As Lyon turned out for the dawn action stations of that day, he reflected that at this rate they would make their objective for the first "hide" that evening. At noon they were off the island of Bengku, approximately 104° 16' E and 0° 22' S.

Lyon directed the others to lower the dinghy and take soundings. The prospect of land and the looming of the final phases of their approach lent speed and dexterity to the men's actions. The boat dropped into the water with a light splash.

Suddenly there was a sharp call from Lyon. The wind had carried to him the distant sound of an aircraft engine. Then he had spotted it, coming in low and fast. He shouted for all hands to cover, except Davidson, already in the boat. He threw down to him a conical native straw hat.

Now all could hear the rapidly increasing drone of the aircraft.

Every man but Lyon was under cover on the *Krait*, while Davidson crouched in the dinghy, his form half-covered by the ragged straw. Lyon was similarly attired. He lounged against the rail as though talking with

Davidson. He knew from the suddenness of the approach that the enemy craft was flying low. Had she spotted them before, and was she closing in for a good look?

The single-engined patrol plane thundered over at one hundred feet. Lyon looked up, as would be natural, and saw the pilot peering down. The stain Lyon was wearing served its purpose.

The Nipponese plane did not circle, but flew into the distance and was gone. Lyon was thankful for the Japanese merchant service flag under which they were sailing.

That night they swung on their hook in the lee of the island. The *Krait*, it had been decided, would off-load them, then cruise off Borneo until time for the pickup again. As discussion ended, one of the men appeared at the scuttle, motioning Lyon topside.

The heavens were bisected by the restless beams of searchlights. Searchlights likely meant guns. They had landed in a distinctly "hot area." Better not unload here.

Later that night they were groping off Panjang. Towering cumulus clouds had broken in a sudden tropical downpour. The *Krait* pitched heavily. Then here and there in the seething murk blurred white lights became visible. Davidson thought they must have lost their way and were actually entering an unsuspected enemy small-boat base. Page called guardedly for Carse. The navigator appeared. Lyon clung firmly to a stanchion and looked inquiringly at him.

"No worry," reassured Carse. "They are fishing pagars. I plotted their positions before. Now I know where we are."

Carse was referring to the pressure lanterns hung by natives at the entrance to their fish traps. Attracted, then blinded by the light, fish would swim into the bamboo yards and be unable to find their way out.

The storm passed. They made Panjang and worked the rest of the night to unload. One party lightened ship, the other transferred stores from the beach and buried or otherwise concealed them. Another small party made a reconnaissance of the island. With the faintest flush in the eastern sky, the *Krait* stood high out of the water. Words were few as Carse leaned over the side and shook Lyon's upraised hand.

"See you on Pompong Island the night of October first or second," said Lyon briefly. "Cheerio."

With hardly a ripple the black little ship slid away from them and pointed her bow toward Borneo, where she would hide. Lyon turned to the others, solemn and quiet. Action was needed. He issued instructions to overhaul all canoe and limpet gear. One party would eliminate all traces of tracks on the beach.

"No need, sir," replied Operative Huston. "It's been done for us."

"You mean natives?" queried Lyon sharply.

"No, sir. Hermit crabs. An army of them. Cleaned off the beach as neatly as if they had brooms."

The men overhauled gear, then rested for a day and a half. After that the canoes were loaded for the first lap of the final journey. The plan called for one more "hide" on Dongas Island, within eight miles of Singapore Harbor entrance. This spot where they were now would be their rear base.

It was nearly dusk. The canoes had been camouflaged with green branches. Lyon looked at his watch. He held up one hand. In two other canoes an arm was raised. They were ready. With him was Operative Huston. Lyon's arm fell. Huston's paddle dipped. From the low banks three clumps of tropical thicket seemed to detach themselves and drift along the shore. They were off.

At that moment came the drone of a power-boat engine.

Lyon's canoe bumped lightly, dipped, recovered, and snuggled up to the bank again. A quick glance assured him that the others also had sensed the danger; without a word between them they had resumed their "hide" stations.

There was silence except for the throb of the engine, intensified as an unlighted patrol boat the size of *Krait* plodded around the point three hundred yards away. They could see the snouts of machine guns and something larger—a one-pounder, perhaps.

The minutes passed.

Lyon relaxed. The patrol had not seen them. For the enemy that would be a costly failure. A few seconds the other way and there would have been no "Jaywick."

They slid off. But now their cards had to be played close to the chest. Three times they froze as other patrols muttered by unseen in the night. Other shadows that they could not define also came and went, together with multiple sounds, until their nerves were tight. To port was a small island. Lyon's decision was that of a leader. They were in no shape to try further for Dongas. They went in.

Two days later on Dongas they found a perfect "hide" and a perfect observation point as well. Eight miles from them Singapore's Kallang airport could be seen perfectly. All during the night hours they observed ship traffic in Keppel Harbor, for there was no black-out in this wartime city. With the greatest of satisfaction they observed a considerable movement of shipping in the main harbor. At no time was there less than a hundred thousand tons. The ships arrived from the east, either singly or in groups, but except for one group, none was escorted. All were heavily laden and proceeded directly to anchorages. Outbound ships were either lightly laden or in ballast. Absence of patrol vessels and the freedom of movement of the heavy ships betrayed the fact that there were no minefields in the harbor. Everything *seemed* right for the attack.

But it was not to be from Dongas. Adverse currents drove back their canoes. Eventually they moved to Subar Island, still closer to Singapore.

Now Lyon considered silently. Bulked blackly against the lights were the hulls of numerous ships. The currents were more favorable. All through the months of training Lyon had been aware of a little hot spot of anticipation somewhere deep inside him and he had always felt that when the time was right he would know it. It was dusk of September 26.

They loaded the limpets.

"Got your 'sweets'?" Page had queried solemnly in his canoe.

Davidson nodded. These "sweets" were composed of cyanide within a rubber pellet. Each operative was thus "protected" against torture, should he be caught—protected by the unassailable defenses of death. One bite on the pellet would do it.

A final check over and they were off.

Paddling heavily, Davidson and Falls made only fair time, for the flood tide was on the starboard side. Pausing only now and then as the searchlights from Blakang Mati came too close, they brought up with the pylons of the Keppel Harbor boom in good time and without being sighted. Better luck yet: the boom gate at the Tanjong Pagar end was open and no boom vessel was in attendance!

Soundlessly their paddles pushed them along toward the dark shapes of two ships tied up against the east wharf. But, large as they appeared to the two in the canoe, they obviously were of a tonnage too small to warrant further attention. And then, almost before they realized it, a large vessel loomed close. She was a fast steam ferry, bound for some point south of Blakang Mati. When she had passed, her navigation lights left the harbor aglow with wriggling serpents of color.

Faintly outlined against the glow of the city's lights were what they sought: ships, several of them, of a size worthy of their best. Davidson held up his right hand, doubled into a fist, and Falls grinned at him in the darkness. Yes, this was it! They had limpets enough for three of them.

At this moment, some distance away, Page and Jones were drifting soundlessly toward a black-sided monster that extended above them into the night. Not a word was spoken, not a paddle was dipped. They dared not make a motion, for on the wharf nearby, moving in a circle of light that must have blinded him and helped them, a Japanese sentry stared out into the black water of Keppel Harbor. The drift slowly carried them farther from him. Now the two men went silently to work.

Major Lyon and Operative Huston had reached their target area at about 2130 hours. Silently they paddled, even the whisper of the water alongside being subdued. Then Lyon indicated with his arm. A red light shone off their port quarter. The quivering ruby streak upon the water was like an inviting path. A red light at anchor meant a tanker.

Silkily the canoe glided toward the beckoning light. The tanker's bulk was a gigantic Goliath to the tiny David approaching her. Now it reared over them. Directly approaching from the stern, they maneuvered one of the primed limpets on the end of its retaining rod toward the propeller shaft housing. Black waters slapped noisily against the towering rudder post. For one moment, as the leechlike limpet was being lowered against the steel housing, it seemed to Lyon to be wrong that with a little device that they could carry easily in their arms they should blast a splendid ship into twisted metal.

The magnetized claws of the limpet snapped themselves upon the plates of the tanker. The canoe drifted off forward toward the engine room—there was no use placing limpets over the oil holds which would only flood—but the engine room would fill; if the ship didn't sink outright, at least she would be crippled, for nothing could save her from settling at the stern. Moving imperceptibly, Lyon shifted his position and steadied the canoe. Huston prepared to slip into the water.

But suddenly he stopped. His body was motionless. He was a pale statue in the night, illuminated by a sky glare in the distance. He was staring upward. Lyon followed Huston's line of vision. What he saw momentarily drained his muscles of power and his brain and will of volition.

Above them, not more than ten feet above their heads, a face was thrust through an opened porthole. The man was watching them intently.

For one second, two or three, the taut tableau persisted in a sort of suspended animation. Then Lyon was conscious of his hands moving. He was actually fixing the primed limpet to the rod. Now he nodded to Huston. Huston must understand that he was to proceed.

Without a backward look Huston slipped into the water. The lengths of the death-charged rod shortened as it went down in Huston's grip beneath the surface.

Lyon's brain pounded with a question: Why didn't that man up there raise the alarm? Why didn't he spray them with bullets?

In his state of tension Lyon jumped when the man above made some muttered query in Japanese.

Lyon was surprised to find himself speaking—or rather it was just a meaningless guttural that came from his throat.

The face withdrew.

Huston surfaced and raised his arms for a lift. Lyon heaved. Huston was in the boat.

Above them there was a flash of light!

But there was no shot. Lyon looked up. He could see a berth light, and, dimly, a white-painted, riveted bulkhead beyond that.

The man must have mistaken them for fishermen and turned on his berth light to read. Or had he gone for help?

Lyon seized his paddle. Huston found his and now the bulk of the ship with her single pale eye from the lighted porthole receded.

Lyon recalled later how the strength went out of his arms. His body seemed to have turned to water. Huston sagged. For some moments they drifted, incapable of movement.

But they had to move. The limpets were set for five-thirty. They'd have to be at Dongas Island twelve miles away. The other canoes ought to be there also. They got control of themselves and bent to it.

They had made several miles when Lyon calculated that it was nearly time, nearly five-thirty. And then came a rolling reverberation that could have been four or five distant explosions shortly spaced. Almost at once came five or six more.

They had done it!

The detonations were followed by the distant rise and fall of a siren alarm. Fighter airplanes droned angrily and were lost among great black smoke pillars above Kallang.

It was months after their return to Australia before confirmation came. In the Bureau we had seen their report of claims. But only confirmed "kills" count in the game of war. Then there was a summons to the Naval Intelligence Office. Fiery Captain A. H. McCollum, who had always "chewed" AIB first then said "yes" to our requests, announced that a report had come across half the world "about your—er—'bluejays.'"

"Jaywicks?"

"Jailbirds! Anyway, they did it. Authentic reports routed from Singapore operatives through to Chunking and from there to London and from London to Washington." He waved his arm dramatically in a general sweep over the world map on the wall. "But they finally told us. Forty-six thousand tons! D'ye hear that? Forty-six thousand tons of shipping your jaybirds got. Six ships, and one of them a ten-thousand-tonner!"

His fist hit the desk.

"That's *joltin'* 'em! Good business. Didn't know what hit 'em! Come again when you want something."

The enemy *really* didn't know what had hit him. Later a captured intelligence document attested to that. It said:

> Singapore shipping espionage has been carried out by natives under European instructions. . . . An enemy espionage affair developed early in the morning of 27th September, 1943, at Singapore. It was commanded by Europeans hiding in the neighborhood of Palai in Jahore. It was carried out by Malayan criminals through a Malayan village chief, and the party was composed of ten or more persons, all of them Malayans. As a result of the raid, six ships of two thousand, five thousand tons (three tankers among them) were sunk by bombs due to a clever plan.

Come again, Captain McCollum had said.

AIB did, a year later to the month.

Perhaps those devoted to the astrologer's art might have said that the stars had not been right for "Rimau" project. "Rimau" was to go into the same area at the same time of year for a similar purpose, but with even better equipment.

It was September 11, 1944, when the British submarine *Porpoise* slipped her moorings at Perth and almost immediately dropped to periscope depth. It would have been difficult for anyone to have guessed her course for even a few yards. The slight wake of the hooded eye soon was lost in the dance of whitecaps.

Her slender steel hull was now pointed northwest, and within it were some familiar faces. There was Lieutenant Colonel Ivor Lyon. There was Lieutenant Commander Davidson. Beside him on a sling bunk, hunched over to accommodate his bulk to the restricted quarters, was Captain Page. Talking quietly to new men of "Rimau" up forward were Operatives Huston and Falls together with F. W. Marsh, also of "Jaywick." On uniforms they had left behind were ribbons denoting British Empire recognition of their successes in "Jaywick."

This, then, was the beginning of "Rimau." What was the end? No one knows for certain—because no one came back.

Some things are definitely known, at least of the early stages of the project.

London had approved the plan, which involved piracy as well as daring commando acts. Through her single eye the *Porpoise* sought a victim, but this one was to be "taken alive." Not a man-of-war nor even a fat Japanese merchantman; this time the quarry was a junk—an oriental tramp of the oriental seas.

Just over "The Line" they encountered her, the *Mustika*. The action was swift and decisive. They replaced her native Malay crew with "Rimau" operatives. From the slatternly vessel now came the sounds of hammering, tearing, sawing. While the *Porpoise* stood watch, the work went on. When it was done, a transfer was accomplished; in snug pockets between the decks secret one-man submarine and radio gear were stored and covered.

It was intended that the mother submarine would tow the junk to the island of Maraps, which would be used as a forward operating base. When this was assured, the *Porpoise* would take the junk's Malay crew and push off on her own normal hunting mission with torpedoes in the China Seas; eventually she would return to Australia and intern the Malays. It was arranged that another submarine, the British *Tantalus*, would come to Marapas and pick up the "Rimaus." In the meantime, the junk *Mustika* would have penetrated to within a few miles of Singapore Harbor, now crowded with warships as well as merchantmen—indeed a rich hunting ground. She would have debouched her one-man submarines for the raid, and after the limpets had been affixed and the baby subs had come back, she would have gathered them in and made for Marapas to await *Tantalus*.

But after the departure of *Porpoise*, the record becomes fragmentary. Under Mott's successor, Colonel H. A. Campbell of the British forces, a one-time Malayan planter who had been "Jaywick's" coordinator, an indefatigable investigation was conducted both during the war and afterward. From captured enemy documents came some references. From newspaper accounts published in Singapore during the Japanese occupation came others. There were discrepancies. At last only one solid fact stared out from the shambles of "Rimau"—not one man of the twenty-two in the party was alive to give his account.

It appeared that the *Mustika* had preceded from Marapas without incident for a short time. Then while in sight of inhabited land she had been intercepted by a launch containing native police. Why they had become suspicious—or if they had actually been suspicious at all—is not known. In any case, those aboard the junk had concluded that the Malayans meant to board them. This not only would have compromised their mission but would have permitted highly secret gear to fall into the hands of the enemy. There had been a tremendous blast of small arms and small cannon fire from the *Mustika*. Reports said that the entire complement of the police boat had been killed.

Apparently, however, Lyon had considered "Rimau" hopelessly compromised, as indeed it proved to be. It is axiomatic under such conditions to break up the party to confuse pursuit and for each pair of men to try to make it back to the rendezvous point by separate ways. One segment of the party was to try to sail the *Mustika* back to Marapas. If they were overhauled, they would open the sea cocks. If that was not fast enough, they were to touch off built-in explosives to sink her without a trace. One account stated that the ship actually had to be destroyed immediately and that all of the segments succeeded in reaching the island of Sole, near Marapas—only to be engaged there by the Japanese in a running fire fight. A postwar report of "SRD"—initials for Campbell's renamed "Services Reconnaissance Department"—stated that Lieutenant Colonel Lyon and Lieutenant H. R. Ross were killed at this time, although enemy newspaper stories insist that Lyon was beheaded in Singapore together with at least nine others. (In this instance, the Japanese used the ceremonial sword normally utilized for executing heroes only; exceptionally laudatory comment was included in the enemy account, which said that the men had been taken in a series of island fights in the vicinity of the "Rimau" base.)

It would appear that following the fight at Sole Island most of the men escaped, some reaching Marapas, where they stayed until November 4, obviously hoping against hope that the *Tantalus* would come in early. (Actually she came in eighteen days later and found traces of occupancy but none of the occupants; neither did she find traces of a battle.)

Nevertheless, there apparently had been an attack and the survivors again had broken up and tried to move generally southward. At least three actually penetrated to a point east of Timor Island. A. L. Sargent was taken prisoner here while another died in a fight with a Chinese. The third was said to have gone insane, leaped from a small boat, and been taken by a shark. At any rate, the enemy listed ten more as having been captured at various times, including F. W. Marsh, who later died of disease. The others were those recorded as having been executed at Singapore July 7, 1945. They were Major R. N. Ingleton, Captain Page, Lieutenants A. L. Sargent and W. G. Carey, Warrant Officer A. Warren, and "other ranks;" D. Gooley, R. Fletcher, G. M. Stewart, Falls, and J. F. Hardy. Lieutenant Commander Davidson apparently had died in the inter-island fighting. The fate of the remaining men remains a mystery. They were Lieutenants J. Riggs and B. Reymond, Warrant Officer J. Willersdorf, and C. B. Cameron, A. Campbell, C. Craft, and D. R. Warne.

Thus the scales, finely balanced, ever impersonal, had swung the other way, and in each instance had swung the maximum in this weighing of irregular war in the Pacific.

Tahoelandang

TAHOELANDANG.
The name was something to roll the tongue over and savor. It was redolent of emerald seas breaking on a white coral beach, of a lazy curl of sulphur mist from a somnolent volcano, of gentle, limpid-eyed brown people—and heavily armed Japanese.

The British operative leaned over a map on the table and tapped a point in the Celebes Sea about midway between the tip of the Helmahera group of islands and the southernmost extension of the Philippines' Mindanao. Tahoelandang was one of the smaller of the great Sangir group.

"Giraffe" project had a mission to penetrate the area for several reasons other than the ever-primary one of obtaining intelligence information. A. E. B. Trappes-Lomax, whose adventures in the British service had taken him into the far parts of the world—including the forbidden land of Tibet to run in British diplomatic cipher systems—enumerated them. A number of penetrations in the Molucca area had been compromised almost before they were launched because of AIB's use of water craft not appropriate to these regions. On Tahoelandang there, were known to be a number of native craftsmen skilled in indigenous-type boatbuilding. They were Sangirese—intelligent, loyal, and, when necessary, natural warriors. We proposed to recruit those willing and smuggle them out to Morotai Island, the advance operations base of AIB in 1945. Also, it was known that some Europeans were still refugees on the island, they should be valuable sources of information. The AIB objectives were sound enough, and now "Giraffe" was given urgent impetus by a request from the United States naval headquarters, down on "The Point" of Morotai, for an AIB demolition party to destroy the wreck of a Ventura bomber that had crashed on the beach near Tahoelandang village before the Japanese might take from it valuable classified instruments.

Over the shimmering waters of the Helmahera Sea off Morotai moved a slim pale-blue ship. That would be a United States destroyer on the prowl. Her presence symbolized the peril to "Giraffe." She was checking native prahus. The innocent appearance of native vessels, occasionally fitted with outriggers, was sometimes belied by a sudden burst of fire from concealed machine guns. The smaller ones might have only a parcel of Japanese aboard; the big two-masted war prahus could and did carry formidable concentrations of fifty or even a hundred special-purpose assault troops. One of these could make short shrift of "Giraffe." Obviously the operation would have to be "hit and run."

A critique of the plan was held. Trappes-Lomax was known to be a good planner and an excellent organizer. He held a major's commission in the United Kingdom forces; he personally would head the expeditions. His second would be Lieutenant Brunnings of the Netherlands Forces Intelligence Service. Several others from the Morotai base would go, also two United States naval ratings from the Point to search the Ventura. Trappes-Lomax hesitated, then with a peculiar motion of his head that might have indicated defiance added: "... and R. K. Hardwick."

Hardwick. "R. K." was sixty-five years old. He was also one of the best operatives in AIB. He had forgotten more tricks of the trade than most of the young chaps had learned in all their extensive training. The years had endowed him with much shrewdness, and, when the occasion demanded it, cunning. The years had influenced him, too, to keep his own counsel. He was a lone-dog type of operator. His connections throughout the Great East were extensive, many of them known only to himself. His pipelines for the acquisition of information were responsive only to his own secret tapping.

In his youth he had "gone native" in Borneo. He had picked up most of the dialects in that part of the world. Clad only in a loincloth, he had learned the jungle lore of his Dyak companions. His skin was burned to a deep mahogany and except for his robust barrel build he was almost indistinguishable from them. He became proficient in the use of native weapons and killed his own game food with the Dyak blowpipe. Likely no white man knew more of Borneo, of the Celebes, the Helmaheras, Dutch New Guinea.

It was some forty years later, when war had come, while at the trading and control post of Merauke on the southern coast of Dutch New Guinea, that he had received a certain report. It had stirred him to the depths and moved him to action once more. He knew of AIB in Australia. He would work far southward to Melbourne and seek further assignments with the Netherlands Forces Intelligence Service section of the Bureau there. Perhaps, just perhaps, there would come the opportunity to write a final–and happier–chapter to the story he had heard, while at the same time performing some venturesome and productive mission for AIB.

None in Melbourne questioned R. K.'s knowledge or his ability. But he *was* sixty-five years old, even though he had the physical ruggedness of jungle teak. One project planner after another passed him by until finally he was smothered in the safe, dark back rooms of the Dutch Intelligence group on Domain Road, Melbourne. There he fretted and fumed, and although faithful to his assigned task of compiling a noteworthy compendium on the country on which we were one day to operate, his amazing energies were dissipated in his murmurings against fate. On several occasions I had been party to his wrathful outbreaks over a glass of GI American beer. We began to plot. Still, it was many, many months later before combined pressures finally extricated him from his "Melbourne Commando" status and he joined us on Morotai.

He had heard of "Giraffe" in the planning stages, as was his due. From that moment, through one device or another, he made himself utterly invaluable to the project. He completely "sold" NEFIS and he completely convinced Trappes-Lomax. I could only agree, but entertained uneasiness at his insistence about this one particular job on Tahoelandang. If he cherished hidden motives, no one knew it, for R. K. was a man to keep his own counsel.

The "Giraffe" plan called for penetration by Catalina amphibian plane. The signal annex to the plan was checked and the portable transceivers were tested with AIB's main control station, whose whistles and shrieks forty yards from my rock, wood, and canvas billet atop coral boulders breasting the Celebes Sea were never stilled day or night. Protective liaison had been established. There was a direct line between AIB Operations Control and headquarters of First Tactical Air Force, RAAF, ensconced somewhat uncertainly on the slope of a muddy hill a few miles to the east. Wing Commander Rose would be standing by. Another line would lead to Captain Humphries of the American Air Sea Rescue Group. Thus AIB had quick access to both combat and rescue units.

Trappes-Lomax was adjured to SOS the base if he "as much as smelled troubl;" his force would be much too small for anything but quick thrust tactics, and, in any event, while this operation was in general line with the mission of "Giraffe," it was perhaps somewhat in advance of what originally had been intended. It was one of those jobs that would be well if it ended well. Trappes-Lomax put it differently: "Perhaps," he had said drily, "we should get the real approval after we get back, what?" (Actually, "Giraffe" and other parallel operations had the approval "in principle" of AIB's new controller, Brigadier K. A. Wills of the Australian forces, whose record in both World War I and II marked him as one of the most astute intelligence officers in the whole Allied military hierarchy—otherwise I should not have considered moving, for Wills was justly renowned for running "a taut ship"; he had replaced Roberts, who at his own request had been shifted to other duties in Melbourne.)

Dawn was no more than a reality before the sky quivered to the twin engines of a Catalina pounding for altitude. Conditions were good and the forecast was favorable. Unless she met enemy fighters—not very likely at this stage of the war—she should put her party down on the beach in two and a half to three hours, depending on winds aloft. There would be some three thousand pounds of equipment and other stores to unload, then she would return to Morotai and await the call for pickup; she could not be risked to sit it out at Tahoelandang without air cover, and air cover would tip the enemy off.

AIB had little real information about the situation in the Sangirs.

It was known that the enemy paid regular calls to Tahoelandang, but it was believed that the rajah there was still pro-ally, as were most of his people; it was important to the embarrassment of Japanese north-south communications that they be kept that way. There was, however, some kind of skulduggery going on at Tahoelandang. It was necessary that the true state of affairs be determined. At this stage of operations, however, there was little to do but wait. The Catalina would break radio silence only with a "squirt" to indicate arrival over the target.

No incident had marked the trip outward. The Catalina circled the gem of an island three times. There was no sign of enemy occupation. Trappes-Lomax signaled for the descent. The "Cat's" engines went to the slow position and her hull hit the water almost opposite a village they had seen from the air, and not far from the half-submerged Ventura. The sea was calm. There was no flaw in the transfer of the party and equipment to the shore by means of a rubber dinghy. The thing had been done in training many times.

But the groups of natives that had been visible from the air now were gone, dissolved, seemingly. Laboriously the party dragged the equipment onto the beach.

"Terrified," Hardwick had offered laconically, burdening himself with one of the cases containing wireless gear which would be his responsibility. "They'll come about; just don't force the matter!"

His knowledge of native behavior was confirmed. One by one the more intrepid exposed themselves—only a head at first—until finally little groups gathered at a distance. No one offered to assist.

"Either mistreatment or someone has put the fear into them about helping anyone but Japanese callers," Hardwick suggested. "They'll come about."

It was decided to establish a wireless position on the beach. Trappes-Lomax directed Sergeant Major Perry, a giant of a man and one of the most courageous of AIB's commando operatives, to "take two men and remain with a submachine gun and the radio. Establish squirt contact with Morotai and be prepared to come on forward at a signal." The rest of the party went to the main village.

The well-kept roadway was indicative of the general neatness and preserved state of the village, which had been well administered by the Dutch. Still unspeaking, the now large gathering of natives fell away from them as they entered the heart of the town, as far as a low white wall which surrounded the former district office.

Posting his small party at the most advantageous points for fire control, the leader and his chosen commandos pushed into the place.

It was empty.

That was rather to be expected in view of the fact that the Astana, or rajah's, palace, which they had passed on the way, and hurriedly examined, bore signs of having been hastily evacuated some time previously. From what information AIB had, it was believed that the rajah was not strong, physically or politically, and had depended mostly on one Manoppo, his assistant, and a man of dubious intentions.

Hardwick pounced upon a battered typewriter which stood upon the former Dutch administrator's table and triumphantly pulled from it a paper. Apparently composed in a hurry by one not too familiar with a typewriter, the note was addressed to the Japanese commander at Siaoe Island, thirty-two miles away. It was an urgent appeal for help to repel the "invaders." Manoppo had had time to type only a part of his name before the commando party approached too close for his courage and safety.

Hardwick snorted: that settled the status of Manoppo's political convictions. They consulted. They would have to act fast and root him out of his hiding place before he could launch a messenger to Siaoe.

They clumped outside. To the surprise of Trappes-Lomax, Hardwick broke into a torrent of native dialect which immediately got results. "Sangirese," he said simply to his commander's look of inquiry. (No one knew for sure how many tongues he spoke fluently.) In no time he had enlisted the aid of a small group of natives who were to act as guides.

"But aren't you coming?" asked Trappes-Lomax.

Hardwick fixed him with pale-blue eyes and, stiffening to attention, asked formal permission to be excused in order that he might attend to a little piece of private business.

Trappes-Lomax hesitated only a moment, for Hardwick was an adventurer after his own heart. He nodded and, to his amusement,

Hardwick immediately dug from someplace a batch of old swords, brown with rust but still serviceable, and forthwith armed a small party, which he energetically charged in Sangirese, and sent them off into the jungle. Then he reported that he had confirmed through the natives that Manoppo had indeed evolved into a brutal petty despot who killed and maimed, and held the population in utter terror. Leaving Hardwick in charge of the beach, Trappes-Lomax moved out.

At Morotai the aerials on the tall palms over the atap receiver shack got only some brief cipher messages to indicate that a moderate quantity of intelligence information had been picked up, some valuable documents secured, and—innocent phrase that it was—"some useful refugees"—would be returning with "Giraffe." Trappes-Lomax would make a reconnaissance of the island before requesting the Catalina to return and pick them all up the next day.

Morotai was quiet except for the tide that lapped in noisily.

The next day dawned fine and clear.

On Tahoelandang, Trappes-Lomax had commenced his reconnaissance to survey the island for possible use later as an operations base. Manoppo had not been found, but apparently no prahu had left the island. Nevertheless, he would make the survey a quick one, for it would be foolhardy to play their luck too far. Hardwick and Captain V. D. ("Big Dave") Prentice, a tremendous Australian commando, were left with the radio operator and the two United States ratings armed with army pistols.

The morning went quietly enough in the beginning. But Hardwick was "nervy." Perhaps it was natural, knowing as all of them did that it was only a matter of time before something would happen. There was another reason, however, and it lay heavily on his conscience: the little party he had armed with sabers had returned, and now they had company in the village.

Nine o'clock, ten o'clock, half-past ten.

He saw the small prahu coming. Four Sangirese leaped ashore and ran to him. Breathlessly they told of the two big war prahus being made ready at Siaoe. Hundreds of armed Japanese soldiers had been standing by waiting to board when the messengers had managed to slip away undetected and make for Tahoelandang with all speed.

Hardwick pondered. He was at the moment only an honorary subaltern in the Netherlands Forces and quite without authority. He turned to his telegraphist.

"Can you encode a message?" he demanded.

The man nodded. He was an operator, not a coder, but he could try.

The message received at Morotai at approximately two o'clock in the afternoon was a headache for the decoding clerks, and the static crashes that heralded the storm making up over the Helmahera Sea had done nothing to make the transmission clearer. But the gist was clear: "attack expected" and "send aircraft soonest."

By now the storm was so close that the static could be heard even on the land line to RAAF. Through it came the reassuring: "Rose here." But after the situation had been outlined and aircraft requested, there followed a silence so prolonged that AIB was certain it had been "given the treatment" by some field switchboard operator. Such was not the case; Rose was not a man given to impulses and the weather had made him doubly thoughtful. Only a Catalina which could be used for a water landing would be required. But "Cats" were no rough-weather birds. While we spoke a second message came in from Tahoelandang: the enemy prahus had taken off from Siaoe.

"I'll send fighters armed with rockets," announced Rose. "That may discourage 'em until we can get the Cats up."

It was the best that could be done. Outside our operations shack the day had gone dark and there was no horizon. But above the roar of the wind came the spattering of aircraft engines as the Beaufighters drove into it. Could they intercept the prahus? Or would the poor visibility conceal them and let them slip into Tahoelandang?

It was with mingled feelings that we received Rose's announcement that not one Catalina but two would be taking off within ten minutes.

"In this?"

"In this. God bless 'em."

Now not only the Tahoelandang party but the crews and their two aircraft were the chips that had been thrown into this tight game for life itself. Radio was trying to raise "Giraffe" party to advise them of the latest developments and to stand by for possible boarding sometime after five o'clock. Radio was not sure whether they had got "Giraffe's" "Roger" because of the lacerated atmospherics but they "believed" they had.

The sense of isolation grew upon us. Even the RAAF base could not be raised on the land line. AIB operations could not be sure that the Catalinas had gone. Normally they would waddle out of their revetments at the bomber strip north of us, whereas the "Beaus" took off from the fighter strip and went over us.

The door opened and borne in by the cascade of water from the storm and his own streaming slicker came Flight Lieutenant Rod Marsland, my personal assistant at the base. His presence was reassuring, this veteran of dozens of adventures, from rescuing parties off New Britain by air to riding an assault tank ashore with the Marines in order to set up an AIB intersection point for radio air warnings. The minutes went by while we paced the operations shack floor.

Five-thirty. Then it was six. And six-thirty.

Radio could raise no one. Marsland again spun the hand cranks on the field telephones. At last one came to life. It was RAAF, but RAAF had no new word. The Cats were on radio silence. Maybe it was the storm. Bomber operations? No word. The fighter strip? No contact.

Seven o'clock.

The RAAF line bell rang. Marsland snatched up the headset. He canted his head to hear better. Suddenly he shouted.

"Roger! Plane 44 radioes that her ETA [estimated time of arrival] at the bomber strip is 1940 hours. Roger." Then he was listening again. "What's that...? You're pulling my leg...."

He cupped the instrument and turned around.

"Kids," he said laconically, "Hardwick radioes that he's got some kids aboard number 44–two women and four children."

We stared at each other. Women and children at an advance operations base of a "hush-hush" outfit?

Rod picked up another telephone and called the Netherlands Indies Civil Administration number. The Queen's government had installed these units as rapidly as a semblance of stability appeared following consolidation by the combat troops in order to reassert Dutch civil authority in their territories. The Dutch administrator at NICA was unbelieving but cooperative. Wisely he asked no questions, for we could not have answered them. He was advised that we would feed the incoming party and then transport them the six miles to NICA by ambulances and trucks, for by now it appeared that not less than thirty-six refugees would be arriving in the second plane due at ten minutes to eight.

It was just after eight o'clock when a long string of lights came blinking among the dripping palm trees and jungle bush at the far end of the AIB compound. The guests were arriving. The field kitchens had been alerted. That would be a good welcome, better than speeches.

Half an hour later Trappes-Lomax reported in. He listed as "Giraffe's" take: (1) quantities of records, police, local, and economic; (2) some twenty thousand guilders of Dutch money that had been kept hidden from the enemy; (3) eight excellent guerrillas to form the nucleus of a guerrilla company; (4) a relative of the rajah from whom Hardwick expected much information; (5) four European children rescued, together with two "worthy" Indonesian nursemaids; (6) a group of European business people loaded with information, who, in any event, could not be left behind, since that had assisted "Giraffe"; (7) the Ventura destroyed after her instruments had been salvaged.

The field phone rang again. It was Rose to say that the Beaufighters had returned without having sighted the prahus, but that they would go out again "at first light." (They did, and "got" more than a hundred Japanese who had landed at Tahoelandang and fortified the place.)

"I suppose," said Trappes-Lomax, "you'll be wanting to see Hardwick."

We did indeed, and it was sometime before dawn that Hardwick finished his narrative. It began with the story he had heard many months previously in Merauke. Before him had lain the hot, turgid Merauke River. To the south was the brilliant blue of the Arafura Sea. Far to the north was the lofty rock spine of the Orange Ranke Mountains. Over those ranges from the north via one of his secret pipelines had come the story.

Murder had been committed by the Japanese invaders of a small, peaceful island. The enemy's much-feared Gestapo, the *Kempei Tai*, had been determined to unearth information relative to the use of wireless receivers by islanders who were listening to Allied broadcasts.

They got their chance. They had taken away this woman, a Hungarian whose missionary-doctor husband had been abducted into internment some weeks before. She was tortured to make her talk about the concealed radio receivers. This went on for six weeks. Near the end it was arranged for her eldest child, Emma, to "visit" her. The child was filled with horror at what she saw. The commander then ordered that the mother be executed by beheading.

Hardwick had demanded to know what had happened to the children. His informant was not sure but believed that in some way or other they had escaped to temporary safety. Emma, Eva, Djoela, and Joseph Cseszke, ranging from thirteen to five years of age, were thought to have been spirited away together with their *mantri verpleegster*, or Dutch-trained Indonesian nurses, to Tahoelandang Island. Hardwick repeated the name until it was part of his consciousness. Many months later he heard it mentioned in the planning stages of "Giraffe." Truly coincidence was more amazing than fiction.

The next day we visited NICA.

The four children were pathetic figures, their Indonesian nurses little better. One of the latter was deformed by torture she had received, but, unlike her mistress, she had been released. All were fever-ridden, ulcerous, and bewildered, this time by kindness. In spite of everything, the alertness of their mentalities had not been dimmed. Even the youngest fluently spoke a mixture of Sangirese and Hungarian.

It occurred to us what a hodgepodge it was: Hungarian children rescued from a South Pacific Sangir island, together with their Indonesian nurses, by a sixty-five-year-old English national who held an honorary commission in the Netherlands East Indies forces!

"NICA is sending them to Australia," said Hardwick. "The first and I believe only European children rescued from the Japanese in the Netherlands East Indies." He blinked. "Aren't they—aren't they—lovely?"

"Are you going to be their father?" Rod asked.

Hardwick shook his head, almost sadly. "No," he answered slowly. "Fortunately for them, because we believe their own father is alive and well, even though a captive. No, it's not for me, I guess," he added, staring into the distance. "Another job or two, and then I guess I'll have had everything."

"Yes, R.K.," Rod agreed. "Even quadruplets. And at your age!" Hardwick's pale-blue eyes were snapping. "What's my age got to do with it, anyway?" he demanded irritably.

Sultan's Ransom

THE WATER-SOAKED DAY HAD POURED itself cheerlessly into a wet night that muffled the pulsating of the engines. Two PT boats moved away from the pier at the Point and breasted the dark waters of the bay. The Navy signal station tower winked a beam of light over them and the shadowy figures on the PT's noted time, positions, and navigational data. The heavy night closed in after them, and they were gone.

It would not be a very long run for these distance-devouring craft, whose stout motors now pumped them through a choppy sea at twenty knots. No light showed. Shortly before 2300 hours--11 p.m.–the motors were cut. On the port side there was the dark bulk of land. This was the island of Hiri, approximately two hundred miles south-southwest of Morotai. The skipper of the leading craft spoke to the leader of the NEFIS-SRD party. He pointed to a break in the beach line. This was the location of the village. The objective of this "snatch" raid was a particularly well-concealed individual who strongly desired to be extricated from self-imposed exile and taken to safety at Morotai.

Still with no show of lights, the rubber boat was inflated and lowered into the sheltered waters. In a moment it was lost among the darker shadows of the land upon the sea. The reconnaissance party had gone ashore. On the PT's deck stood R. K. Hardwick, on his second AIB Morotai-controlled job. Presently a torch on shore blinked the prearranged "all clear." Hardwick shifted the gear from the crowded little deck into a second dinghy, then got into it himself. The PT's stood by until another signal indicated that the second dinghy had made the beach, then departed.

Hardwick later reported that it had been considered advisable to lose no time in contacting the "objective," who was the Sultan of Ternate, the Malay ruler in the days of Dutch sovereignty over the area. The native peoples were trusting the Allies to protect him; his capture by the Japanese would alienate countless Malays and render AIB's position in the Moluccas and the Celebes precarious. It was the mission of "Opossum" party to evacuate him and his immediate family before the enemy could effect his capture.

Accordingly now one of the native operatives slipped into the shadows of the thick tropical growth leading to the beach. His ultimate destination was the island of Ternate, due south, across one mile of dark water. In his mouth he carried a tiny phial. Inside the phial was a communication written on the thinnest tissue paper. Should the messenger be captured, he had only to swallow the tube and explain that he was ill and was endeavoring to find some medical help for his throbbing head.

Lieutenant "B" consulted with Hardwick. It was B's first assignment as an SRD component leader in this combined Dutch-SRD job and he frankly sought the dictates of experience. Would it be advisable to penetrate deep inland or establish a small defense perimeter on the beach? In the absence of detailed up-to-date intelligence information about the island, Hardwick suggested they move inland. The senior Dutch officer present, the actual ranking leader, concurred and they began preparations. Sentries were posted and a pair of submachine guns were assigned to the beach approaches. A box of hand grenades was broached and some of the contents placed in most accessible locations.

The night passed without complications, and when dawn broke the party looked upon a scene of great beauty. Before them lay a cobalt sea stitched whitely with coral reefs, and above them reared tall, rose-tinted mountains. But the mission left little time for contemplating this; the "Opossums" were busy with breaking camp and packing equipment for the next move. They marched overland in a southeasterly direction, and presently arrived at the village of Togolabe, on the island of Hiri.

Now there was a break, and all eyes feasted briefly upon the beauty of stately Ternate Peak, just across the strait, its lofty cone lifting gracefully from a perfectly turned base. In the morning's freshness they could discern the nutmeg gardens clinging halfway up the sloping sides of the mountain, softening its lines with their brown carpets. The peak *was* the island, a land fought over by Spanish, Portuguese, Dutch, and British, passing into the sovereignty of the Dutch East India Company in the seventeenth century, when it reached a splendor never since attained.

The party turned again to the job. Nearby there was another village, called Dorara-Isa, the only disloyal village on the island. This one likely would be a troublemaker. The commandos entered quietly.

Before the inhabitants were aware of what was happening, the male population, consisting of some seventy men, had been rounded up. In the center of the gathering, simmering with hostility and savage intentions, was the Machimo, known throughout the Ternate group as a pro-Japanese with a record of much cruelty. He was securely and none-too-gently bound. On each side of him were bound his two thin-lipped henchmen, island versions of the gestapo, whose hands were known to have been reddened with murder. So far, luck had favored the raiding party. It was thought that with the three hostile men secured no word of the raid would reach the Japanese garrisons, only a mile across the water, so much had the experience of Tahoelandang taught AIB! A strict guard was posted and silence enforced.

Hardwick tried to rest in preparation for what well could be an arduous night. But sleep was an uneasy bird that perched only momentarily between restless flights across consciousness. It was no use, it only clouded his faculties. He arose and for the twentieth time cautiously peered across the bright blue water to Ternate. There was no sign.

Came noontime. He was too fidgety to eat and went back to the lookout.

At last, sometime after three, he spotted the tiny dot far out, but coming in at an amazing pace.

The native swimmer lifted his gleaming brown body out of the sea and came into the foliage. From his mouth he took the same phial. An amazing amount of information was concentrated on the tiny paper: prahus lay concealed on the Ternate shore. All was ready. But there were problems in the shape of "traitors and spies" on every hand. The Sultan would have to be sure. And then he would embark not only himself and family, but a retinue of the faithful, for to leave them would be to consign them to torture and death. He would make the passage in the nighttime and come ashore sometime after dawn tomorrow.

That would be Tuesday. The pickup by PT boats was not scheduled until Wednesday. There would be long hours between the present and the next dawn, and then a whole day and night after that. Any number of things could happen in any one of those hours. Hardwick's mind went back to Tahoelandang and a ripple of uneasiness went through him. They should not wait.

With some hesitation Hardwick approached the senior Dutch officer. The grizzled commando was aware that he had skated on thin ice in taking things into his own hands at Tahoelandang, but events had proved him justified. Still, command was command, and he was only an honorary subaltern. Furthermore, he had detected a coldness on the part of the Hollander since "B" had asked for his opinion rather than referring immediately to the Dutch superior. Hardwick's mobile face creased into signs of impatience. Damned silly business when the voice of experience had to bow to the voice of command merely because it was command. He would put it up to the chap, anyway.

Afterward he confessed to an inability to decide whether it had been the wise thing. He got a rejection, not sharp, it was true, but a rejection. Perhaps if he had not mentioned it, the Dutch officer might have ordered it himself. At any rate, he based his refusal to call Morotai on the thesis that it was unwise to ask for an alteration in plans once they had been established, especially when they involved, as they did operations by a sister service, in this case the American Navy. They would maintain the strictest lookout, but they would wait.

Hardwick was irritated with himself. Secretly he wondered if he was getting too old for this sort of thing, after all. Damn! From concealment he studied the enemy-occupied island a mile away.

Everything seemed peaceful there, lost in tropical laze and wellbeing.

Next morning he felt more than ever that his apprehensions had sprung from a brain that had seen more than its share of fire and brimstone in this world. Word had come that a small convoy of prahus had arrived at another part of the island and that the Sultan and his retinue were then coming ashore, intact and happy. He plodded to the beach.

The scene was dramatic. Already many hundreds of natives, more than the "Opossums" had estimated were in the entire area, blackened the white, stony beach. As the small, dark figure of the Sultan strode ashore, whole groups rushed forward to kiss his feet.

Hardwick found himself marveling at his own reaction to this: it was one of approval. There was nothing of servility in what he was seeing; rather there was a peculiar dignity—a reverential expression of loyalty and sympathy toward the ruler. A legacy of the old days, and the flight of the decades and even centuries had not altered in the slightest the customs of these simple people.

The little man with the sad eyes spoke gently. A native ran lightly to him and dropping upon one knee assumed an attitude similar to one praying. Palms were together, thumbs level with the nose. The Sultan's lips moved. The native before him remained unmoving until dismissed. Obviously various native dialects were employed. Yet all of them the Sultan enunciated with equal facility. Later, when approached by the Dutch officers, he conversed with them in their native language, and still later in French. With Hardwick he conversed in excellent English.

Exhausted by his long exile in the heart of the island of Ternate, he explained that he had put reliance on drugs to get rest and sleep. The strain had taken its toll, and his health no longer was robust. But Hardwick could see that the spirit burned steadily in the dark eyes. With him in the midnight escape from his mountain residence had been his two wives, one of them in an advanced state of pregnancy, but they had made the precarious descent down the mountainside with slight assistance. More than one pair of commando eyes surveyed with approval the third member of the royal party— a truly beautiful girl of fourteen, who might easily have passed for seventeen. She was Rene, the Sultan's daughter. Fair in coloring, with aristocratic Malayan features, a lively disposition, and charming, frank manners, she was clad in revealing white shorts. She would have found equally approving eyes at Miami, Hollywood, or the Riviera.

The royal party, numbering some thirty in all, was installed in a firmly-built house previously selected for it by Lieutenant "B's" scouts. For the rest of the day the Sultan received idolizing subjects.

But back of this pomp of royalty there was a tension that stemmed from no less than the Sultan himself. He communicated his views to Lieutenant "B."

An attack, he said, without show of emotion, might be expected late that day, or surely the following morning. That was all. There was no hint of panic, none of urgency.

Perhaps it was this that misled the youthful Lieutenant "B" as well as the Dutch officers. Possibly in the characteristic Malay lack of demonstration they had taken too much assurance. But now the pent-up unease in Hardwick broke out. This time the rebuff was sharp and decisive. Thereafter Hardwick would refrain from disturbing carefully established plans which so far had operated with smoothness and precision. After all, the day was well advanced now, the leader pointed out, and he reasoned that if the attack were coming, there should have been some indications of it. Lookouts posted on high country had seen no signs of approaching trouble via the sea. Early in the morning the PT's would arrive, and anything less than a destroyer that might attempt to stand before those waspish craft with their impressive armament would not live long enough to rectify the mistake. The Dutch officer turned away abruptly. Harwick stared after him. "Sheer madness," he muttered, laying about coldly with his pale eyes. "Not only would a night attack wipe us up, but it would bring death to the Sultan, his household, and no one can say how many hundreds of these splendid, loyal people who already have suffered too much from a white man's war."

He was aware that Lieutenant "B" had heard him. Casting caution to the winds, he pleaded his case.

The Australian officer listened. His brow was creased. But there was the matter of Dutch sovereignty as well as seniority. Lieutenant "B" remained silent, if thoughtful, and took no action. Thus he sealed his own fate.

The sun sank behind mountainous clouds marching gigantically across the western rim of the world. The heat waved up in unseen layers.

Just before a premature darkness without dusk settled down upon the little island Hardwick personally inspected the position. He stumped his way around the little perimeter that had been established. Everything, such as it was—pitifully inadequate to stop an attack of real dimensions—was in place. He selected a hut near the point where the Bren gun had been mounted to cover approaches from the beach.

But sleep was long in coming. Restless and tense at the slightest sound, it was sheer relief to him sometime in the night to hear the fall of heavy raindrops, thudding like lead pellets through the jungle. With the rain came the wind, almost a full gale. The palm fronds streamed out horizontally from the fury of the storm. There'd be no attack during this. And although the storm was too intense to last for more than a few hours, the seas would be dangerously heavy for some time to come. Perhaps nature was protecting them?

At that moment the local enemy commander on Ternate was energetically trying to arrange for the embarkation of an assault group to cross to Hiri. The native boatman had demurred, pointing to the towering waves. There could be no denying the risk in those seas, but it was a decreasing risk as the tropical storm abated. Sooner or later they would have to agree.

Dawn was just breaking when Hardwick heard a startled shout followed by the crack of a carbine. Two more shots. Then an irregular rattling of musketry from various points.

Hardwick was already out of the hut, shouting for two commandos to stand by the Sultan's party. He lunged for the nearest cover and peered along the beach.

Thank God no enemy had actually made the landing so far as he could see. But there were swiftly moving prahus on the surface of the dawn-swept sea.

The men at the Bren gun were working frantically to uproot it. By sheer man power they wrenched up the heavy piece and hustled it along the beach to where it might bear on the closest prahu.

If Lieutenant "B" had been uncertain before, he was in his element now. Everywhere at once, he was shouting for two men to stay with the radio and cover it against all comers. To the machine gunners he shouted directions to "get them in the water and keep them there."

Hardwick approved. "B" knew that if the enemy landed, the dispersed nature of their own guard easily could set them to firing into each other as well as the enemy.

From some point down the beach came a rattle of gunfire. It was cut over at once by the hammering of the Bren gun. Quick geysers spouted in ranks close to the leading prahus. But momentary bursts of spray at appalling distances from the enemy craft spoke eloquently of the kind of marksmanship possessed by the native levies down the coast; these had been armed and given "dry run" practice with the United States carbine— a fine, effective weapon in the right hands.

Against this handicap was matched the handicap of the enemy in being caught on the open water. Splendidly trained, beautifully armed, and fanatically brave, the assault party drove in despite the heavy count the Bren was taking among them.

From the cover of a palm-tree bole Hardwick was working his carbine with slow, deadly precision. From someplace beyond the Bren he heard the sub-machine guns break into a savage riveting. The sea boiled with bullets, yet they came on.

Fewer now. But they kept coming.

Hardwick drew beads on successive figures as they stumbled out of the water and flung themselves onto the sand, then inched to cover behind logs and rocks. Some rose momentarily to lunge for better cover farther in.

With an eye and hand as steady as if he were doing range practice, Lieutenant "B" spun one of them around with a perfectly placed shoulder shot and brought the Japanese low. The man fell writhing among the white stones of the beach. This was what Lieutenant "B" wanted—a "winged" but live Japanese prisoner who could be made to give information about the Japanese north-south communication lines.

"Don't go near!" Hardwick shouted. He ran out from the tree bole. "They're dangerous that way. *Watch him!*"

But Lieutenant "B" was rushing forward. Apparently he was determined either to take the enemy alive or polish him off.

Hardwick saw the enemy soldier make a swift motion, pivot on his elbow, and fire his rifle at point-blank range.

Lieutenant "B" pitched forward over the quivering body of his enemy. Two bullets from somewhere ripped into the Japanese. Both men twitched convulsively, then were quiet.

Hardwick felt hot and sick. He turned to the fight again—and immediately found himself the helpless witness to another tragedy.

Lithe Private "H," Australian machine gunner, taking advantage of a lull in the fighting—a lull caused apparently by the fact that the enemy party was all but accounted for—had suddenly placed his gun on the beach, stripped off his few clothes, and plunged into the sea, swimming strongly.

His objective was a war canoe. Unquestionably the craft would have been valuable to AIB because bona-fide native boats of the type utilized by the enemy were of utmost use in "sneak" parties; they were able to penetrate deeply without suspicion and detection.

The little white fountains still were leaping about the enemy prahus, and particularly where several enemy swimmers bobbed beyond the abandoned craft. But Private "H" seemed to be endowed with a protective charm.

Now "H" was very close. Hardwick lost sight of him for a moment as he bobbed under the stern of the war canoe, then saw him clambering aboard.

A ragged cheer came from the beach. It was premature.

There was a spatter of carbine fire from the location farther down the beach where some of the native levy had been exhibiting the wildest shooting. But this time, of all times, their aim suddenly took on fatal accuracy. Apparently they had mistaken the suddenly exposed body of the little commando for a Japanese.

Private "H" dropped, a red streak painting its way quickly down his splendid young body.

As though in revenge upon an enemy that had not committed the act, the beach party exploded into a fury of fire. The water boiled about the canoe and the prahus. In seconds four swimmers threw up their arms and disappeared beneath the surface. Not only were the enemy soldiers being cut down, but also the native crews of the prahus that had brought them. It was short, bloody work.

Over their dying heads the ether was carrying the emergency appeal from "Opossum's" portable radio transmitter to Morotai.

ATTACK. AIB CASUALTIES. RUSH HELP AND PT BOATS.

What premonitory urging had motivated the PT commander at the Point to have his crews at the alert? Hardly had the emergency call been transmitted on the land wire to the Point than the PT's churned out and away.

But it would be several hours before they could raise the little island. The enemy had only one mile to go in order to reinforce his routed attacking party. More than fifty veteran Japanese naval personnel were on Ternate and with glasses they doubtless had observed the action.

It was learned later that this second force then had been rushing about to arrange for embarkation. The nature of their delay never was determined. But the god of war was being impartial that morning. This heavily-armed force never got under way. Had it succeeded, nothing could have saved the situation. Much of AIB's ammunition was gone for both the Bren and the sub-machine guns, nor was there very much left for the carbines.

Once more AIB made a tense, urgent appeal to the First Tactical Air Force, RAAF. In a matter of minutes, the Spitfires were singing the characteristic even, monotonous whine of their in-line cylinders as they swooped low off the fighter strip, thundered over our radio masts, and banked steeply westward for Hiri.

There, two canoes were putting out from the beach. One was making for the stricken commando. The other was being rowed furiously on past, straight toward the enemy-held Ternate coast.

The reason for the second move was not immediately clear. Then it was explained that keen-eyed natives of the Hiri levy had detected one more tiny dot, far over near Ternate. Whether they offered the escaping enemy an opportunity for surrender was never known; no court of inquiry sat upon them when they returned to tell of shooting the man dead in the water, just as he had all but covered the mile of sea to safety.

He was the Japanese chief of the *Kempei Tai* in Ternate.

Not all the enemy had been cut down in the water. Hardwick suddenly discovered himself outflanked—by one!

He was hardly able to believe his eyes when he happened to turn and saw the enemy officer at his back. In an instant he had put the bole of the tree between himself and the man. Yet he was too close to fire without revealing himself. With inner rage he realized that he was in a bad spot for his inattention. And then it came to him that the Japanese was in an even tighter spot, for being intent upon Hardwick, he had not covered his own rear—and down there was the beach party!

There was a shout from the beach. The Japanese spun around— and froze. Then, with a bland smile, he dropped his pistol.

But suddenly he dropped to the ground, his outstretched hand groping for a broached box of hand grenades.

There were two shots. His legs straightened out. He was dead.

Apparently the entire raiding party was accounted for, except one. Almost at the moment of this survey, the Hiri native levies thrust him forward.

With an exclamation of delight, the Dutch officers greeted the coming of the sullen captive. Long and black was the list of his criminal offenses against both the Dutch and the helpless natives. The "trial" was a model of efficiency. In their turn, the Dutch commandos sentenced him to be shot. Then he was shouldered before the Sultan. The sad-eyed man who had been forced to see his people mistreated and tortured by the ruffian nodded and jerked his head for them to take him away. He was led off by native levies, those whom he had so badly treated in past days. The jungle muffled the reports of two shots.

There was a moment of silence. Then each man looked around at the others all about him. The fight was over.

Quietly they turned to the sad task of collecting and burying the dead, except for the AIB men, who would be returned to Morotai—if any were fated to return.

Half an hour was consumed in the work of clearing and gathering their scattered equipment.

Suddenly there was an exclamation from one of the Dutch officers. What he was staring at was a thing that simply could not be!

The "executed" man, his face and clothing spattered with blood, came running toward them, imploring mercy—and crying wildly his protestations of loyalty to Dutch, Americans, and Australians alike.

It was grisly business. This time the firing squad made certain.

Now they took grim inventory. In favor of the defenders was the capture of a Japanese light machine gun, together with ammunition, from one of the prahus. They had the advantage of defense. Against that it had to be presumed that if the enemy force was capably handled, it doubtless would be split to enable investment of Hiri from several points simultaneously. Ammunition for the Bren and the Thompsons would be gone after a few sustained bursts. The Japanese gun could traverse only a limited front. That would leave carbines. It had been proved that the native riflemen were as dangerous to friend as to foe. ... . The sober estimate of the situation left no doubt but that the best that could be hoped for would be a delaying action sufficiently prolonged to enable some kind of help to come from Morotai.

It was a thin hope. The voices of the conferees trailed into silence.

In that moment the first faint pulsing came to them. The thought in every man's mind was that the enemy had motor pinnaces as well as prahus. The senior Dutch officer opened his mouth to give the order for action stations. But from Hardwick came the bull-like roar:

"Spitties! God bless 'em!"

From the first faint pulse the sound swiftly built up to a high-pitched drone. The fighters whined in and dipped to indicate that their pilots had taken in the situation. They had wing tanks. That meant extended endurance over the target. On the beach reaction set in. Some had "the quivers." Some laughed too loud. Others just stared, or sank down on the sand and were silent.

At midday the PT's drove in.

To Hardwick fell the task of embarking the royal family. It was no easy matter, considering the difference of level of the small native craft and the PT's, and especially in view of the condition of one of the Sultan's wives. There was no time for ceremony. Where a younger man might have hesitated over the proprieties, the veteran Hardwick dismissed graciousness as being a poor substitute for effectiveness in this moment of necessity, and in a mighty heave "boosted" his royal passengers upward.

Then he looked around for the adorable Rene. But she required no "boost" from below. He watched her leap easily to the naval deck with all the lightness and grace of a fawn. Hardwick snorted. Life was like that.

Without further delay the little convoy made for home with all speed, the living and the dead crowded together in the little ships.

As they approached the naval base, the whole company came to attention on decks. The American flags which had been standing stiffly to the wind of their passage fluttered to half-mast. There was a brief, tight interval of silence. The flags were raised to the peak again.

Some days later there arrived, through the bewhiskered Van der Plas, Queen Wilhelmina's personal representative in Australia, an official letter from Dr. van Mook, the vice-governor of the Netherlands East Indies. It expressed mourning for the dead and then, in warmest terms, spoke in appreciation for the "Opossum" exploit. It was read in an impressive ceremony just before Hardwick was to depart for the field headquarters of the Australian Imperial Military Forces. He was to be accorded a commission as major in the Australian Army—at the age of sixty-five. It would constitute an eloquent testimony to his bravery, experience, usefulness, and— perseverance.

"Yes, it's very acceptable." he acknowledged, almost stiffly, referring to the vice-governor's missive. "We've already obtained more useful intelligence from the Sultan than we ever hoped for, even knowing he was loyal to us and anxious to help our cause."

He took a hitch to his khaki breeches.

"My friend," he said solemnly, "you should have heard the commendation I received from the United States Navy."

Really? A commendation from the United States Navy. We had not heard of that. What was it for?

"For?" said Hardwick. "For the thoroughly efficient—quite ungentlemanly—and entirely capable manner in which I handled the royal buttocks in heaving the encumbered wife of the Sultan aboard during a period of great peril."

With that he stomped off, enormously dignified, splendidly disdainful. ...

Wunung Plantation, Jacquinot Bay, New Britain. December 1944. Officers checking map reports at HQ Allied Intelligence Bureau. From left to right: Lieutenant C.K. Johnson, Allied Intelligence Bureau; Flight-Lieutenant H.R. Koch, RAAF; Lt. C.C. Eling, HQ 6th Infantry Brigade; Captain J.W. Mott, HQ 1st Army.

Tinian, Bougainville. July 1945. Sergeant Lea, in charge of Allied Intelligence Bureau native soldiers.

Part 6
FINALE

Wunung, Jacquinot Bay, New Britain, August 1945. Entrance to Allied Intelligence Bureau camp area.

Tamkaidan, New Britain. January 1945. Lt. J. Sampson, Allied Intelligence Bureau, leading a patrol through dense jungle behind enemy lines.

The Unspoken

THE GLASS RUNS OUT. Yet the story is not half told. There is the saga of the unquenchable determination of Lieutenant B. Fairfax-Ross to lead his extensive party out of a hopeless position in the rugged area inland of the Rai coast after the enemy had been able to control nearly all of the natives of the area early in 1943. From February until June the party defied every effort to exterminate it, the innate stubbornness of it being exemplified by Fairfax-Ross himself, who, already weakened by lack of food, was wounded twice March 30 in a fight with a strong enemy force— including a bayonet thrust in his right knee, later developed pneumonia, and still refused to throw in the sponge, lived to walk endless miles to the air-lift point at Bena Bena June 20.

There are the chuckles about the time on the Rai coast when American bombers made it too near a thing and the Australian Coast Watchers had to hit the slit trenches—"Blue" Harris discovering that the man he was sharing his tiny trench with was not another Australian at all but an enemy soldier. And both of them leaping out in different directions simultaneously, bombs or no bombs!

The roster would range from beer-loving Australian H. R. Koch, who was the "pilot's pilot" for the "biscuit bomber" runs out of Nadzab to supply isolated Watchers on dark, dangerous coasts, to scholarly William Harwood, the Tasmanian schoolteacher, who proved to be a wizard at creating cipher systems to keep the enemy in ignorance. It would include Big Tony Gluth, and equally big Dave Prentice—for them, no commando job too tough, too dangerous.

There would have to be a full book on the "200 Flight" of bombers assigned to AIB. In addition to dozens of other missions, they made more than one hundred and fifty penetration sorties in one hundred and fifty days in 1945 when the Ninth Australian Division was depending on AIB for most of its advance intelligence before making the Borneo strike for taking back the rich oil fields. Their shortest flights were eleven hours; their long ones would run fourteen hours. Three of them went down with their crews. They never were heard of again. The "Two Hundreds" did a magnificent job.

There also is the story of the Hadjis. With the permission of King Ibn Saud of Saudi Arabia, AIB was able to bring them out from Jeddah and infiltrate them into dangerous lands where a religious war might serve Allied interests. Some lived to bring back information. Most did not.

The story of the operations on Timor alone fill a volume of AIB records. It was a great pity that SWPA simply was in no position to capitalize on the work of the initial Mott Section agents. But things were too pressing elsewhere. Party after party disappeared on Timor. Signals continued to come out from one penetration group, but the authenticators were missing. AIB could not afford to ignore them altogether; if they were sending under duress, at least they were alive, so periodically the "200 Flight" made drops, at great risk: one was lost doing it. At the end of the war an insolent message came out of Timor from the set known to be the SRD field transmitter that had been sending the suspicious signals. It thanked AIB for the supplies. It was signed "Nippon Army."

There were also the unsung heroes of the base, the people who made the field parties possible. In the United States group there was Bobb B. Glenn, AIB's chief supply officer in the early days. Eventually a major, Bobb earned it. With him, later, was tall, likable Charlie Wilcox—who got thrown in the mud of Morotai by the Australians for daring to be promoted to major. Australians are like that. On Morotai there was bright-eyed Morris Israel and iron-haired William F. Hill, Australians who kept communications going through storm and clear.

So many stories yet untold, so many names yet unspoken....

Finale

SLEEP WAS A FUGITIVE THING, PURSUED MERCILESSLY by thoughts always racing more madly. Under the jungle trees, leaning to the night wind, the atap radio shack was leaking sound through every crevice. Exciting sounds. No longer in cipher, it was now in clear language, for the news was too full of portent.

The Allied capitals of the world were waiting upon the acknowledgment of grim, bowed, defeated leaders of a nation in Tokyo.

It was no use to try any longer to keep to one's bunk. Through the night, softening into the dawn of The Day, one walked a solitary way down the trail past the anti-aircraft battery toward the fighter strip that had been one's track of meditation so many days and nights on Morotai.

The light was pale, sourceless. There were the silent nests of thin, wasp-shaped fighters, then the roomier homes of the Beauforts, their stark airscrews appearing vaguely even larger than their real out-of-proportion size; the bulbous cargo planes, silent and shadowy. Beyond was the beach, where was born The Day.

The amber footlights of dawn silently pushed the night upward. They became stronger until they were expanding girders of light that supported the still-sleeping sky. Against this immense glow waiting coco palms stood tall, sharp, and motionless. Beneath them the upthrust noses of deserted transport planes were black silhouettes on invisible wheels, while on the horizon cloud banks waited in unmoving procession, purple-clad midwives of The Day. . . .

It came that evening, after a day of rumor, counter-rumor, and nerve-stretching uncertainty.

War's end.

From the naval base far down on the Point the tracers lazed out and over the sea like colored water drops from an upturned garden hose. They floated effortlessly and slowly across the arc of a momentary existence, terminating in fraction of darkness and a final tiny exclamation point of white light.

Even war's engines could produce beauty. Yet each of them, somehow, symbolized across the dark graph of the heavens the flight of a human life that they themselves had extinguished in midflight.

Phillips, "Blue" Harris, noble Butteris. Stavermann, Brocx, and the Indonesian who sought him. Olander, fine Lieutenant "B"– and little Private "H." More, too. ... Many more.

Slowly consciousness yielded to something else in the Morotai night. It was music. The radio receivers had ceased their frantic whistling, questing for intelligence. The weary operators had turned onto something else. Quietly it came from some distant unnamed broadcast transmitter, in stately measures, deep in slow benediction. ...

"Nearer, My God, to Thee."

Was it true for us, the living, as well as for them, the dead?

It had to be, for if war could produce no other good thing, it had to produce faith; only the mortal and the works of man could be destroyed by war.

www.ingramcontent.com/pod-product-compliance
Lightning Source LLC
Chambersburg PA
CBHW042111120526
44592CB00042B/2694